Water Management Addressing Societal and Climate Change Challenges

In Focus – Special Book Series

Water Management Addressing Societal and Climate Change Challenges

Editors

H. Ratnaweera, T. Sætersdal, S. B. Weerakoon and F. M. Mutua

IWA PUBLISHING

Published by **IWA Publishing**
Unit 104–105, Export Building
1 Clove Crescent
London E14 2BA, UK
Telephone: +44 (0)20 7654 5500
Fax: +44 (0)20 7654 5555
Email: publications@iwap.co.uk
Web: www.iwapublishing.com

First published 2022

British Library Cataloguing in Publication Data
A CIP catalogue record for this book is available from the British Library

ISBN: 9781789063424

Contents

© 2022 The Authors Water Management Addressing Societal and Climate Change Challenges doi: 10.2166/wcc.2022.001

Editorial: Water management addressing societal and climate change challenges

Water is the primary medium through which climate change influences the earth's ecosystem and thus people's livelihoods and well-being (IPCC 2014). Globally, 83% of all disasters have been caused by extreme weather and climate-related events, such as floods, storms, and heatwaves, over the past decade (IFRC 2020). Together, these disasters have killed more than 410,000 people and affected a staggering 1.7 billion people in different parts of the world (IFRC 2020). With a 50% increase in the global population and a 30% increase in the urban population from 2000 to 2050, the world will continue to face enormous challenges in most contexts. In many cases, poor people and developing countries are most likely to be affected.

The impacts of climate change will make the situation worse by creating extreme conditions with catastrophic outcomes. We may anticipate technological, economic, and sociological impacts resulting from climate change on water resources and processes.

The global average temperature has been increasing at a rate of 0.08 °C per decade since 1880. It has more than doubled since 1981 to 0.18 °C per decade (IPCC 2014). Recent studies predict a 5–15% rise in precipitation per centigrade degree increase in air temperature (Pfahl *et al.* 2017). Since 1901, global precipitation has increased at an average of 2.5 mm per decade. However, much higher increases and decreases in precipitation have been observed within continents, regions, and even countries.

In Europe, 200–400 mm per hour of rain has been observed, leading to devastating floods. They result in damage to infrastructure, water treatment processes, and increased pollution transport to recipients. While investing in separate sewers and stormwater pipes has started in many cities, this process is costly and slow.

Although rainwater is low in or free from pathogens, when rainfall increases runoff on the ground, it leads to increased microbiological pollution in raw water sources. This results from the mobilisation of pathogens in watershed discharges from centralised or decentralised wastewater systems and floods. The relative increase in *E. coli* bacteria could be several thousand times higher, while the increase in faecal streptococcus could be several tens of thousands of times higher (Kistemann *et al.* 2002). If the raw water source/catchment has pollution sources (environmental toxins, pesticides, heavy metals, etc.), this may increase transport to drinking water utilities.

Subsequently, this issue poses health risks associated with drinking water. Utilities will have to increase disinfection capacities in order to maintain adequate water quality. Increased turbidity and suspended solids in raw water also require increased particle removal processes, as they may interfere with disinfection processes.

Colour or natural organic matter (NOM) in raw water sources is common in many countries. It may produce harmful by-products during obligatory disinfection during water purification. A significant increase in NOM is observed in many sources, where climate change is a leading cause (Kritzberg *et al.* 2020). This increase requires removal processes to be strengthened with the adaptation of disinfection processes (Kastl *et al.* 2016; Valdivia-Garcia *et al.* 2019).

With less than 0.3% of all water available as fresh water for human use and unevenly distributed, many regions in the world face physical and economic water scarcity. At present, 47% of the global population, or 3.6 billion people, live in areas that suffer from water scarcity at least one month each year, which is expected to increase to 52% by 2050 (Mekonnen & Hoekstra 2016). A decrease in precipitation increases water scarcity challenges even in non-arid areas, where the demand already does not meet the supply. The droughts and reduced precipitation will also increase the use of severely limited groundwater, resulting in drying up or intrusion of saltwater in coastal belts. Saltwater intrusion is often a non-reversible process or may take many years to reverse. It will further worsen water scarcity, as the population in whole areas will suddenly lose access to fresh water.

By the late 21st century, the global land area and population facing extreme droughts could more than double, increasing from 3% in 1976–2005 to 7–8% (Pokhrel *et al.* 2021). Increasing temperatures can exacerbate existing drought conditions and impacts (ESCAP 2021).

The global mean sea level has risen about 19 cm since 1880 at an accelerating rate (Church & White 2011). From 2018 to 2019, the global sea level rose by 6.1 mm. Besides other natural forgings, the main reason is greenhouse effects, leading to a global temperature increase that melts ice layers and glaciers. Even if the world follows a low greenhouse gas pathway, the global sea level will likely rise by at least 30 cm by the year 2100, compared with the increase in 2000. However, the rising rate is far from uniform between continents and regions (WMO 2021). Economic projections against climate models report a high magnitude of physical risks that 105 countries will experience in the next three decades (MGI 2020).

Climate change may also create positive impacts; there may be significant overall benefits to health and development in adapting to climate change. For example, efforts to adapt to climate change would create a stimulus to aim directly for higher service levels for those currently unserved. Increased awareness of water conservation measures may lead to less water wastage (BBC 2021). Concerns about adapting to climate change create substantial pressure to rationalise the choice of technologies for delivering sustainable and effective services (WHO 2010).

Sound water usage with reduced water consumption began in some cities several decades ago. Wastewater does not necessarily need to be reused as drinking water, but its use for industrial and agricultural purposes will help address the challenges in the drinking water supply. Although water reuse has also become a significant focus in Europe, it still has a long way to go. Systematic water reuse in Asia and Africa is seldom practised, while sporadic use of greywater for agriculture is done. However, domestic greywater reuse in agriculture must be carried out with caution to avoid food contamination.

Recent warnings from the UN on the *new pandemic* concerning Sustainable Development Goal (SDG) 6 – access to clean drinking water for all – have been expressed in a recently published report (UNDRR 2021). The importance of good water management is no longer something that concerns only the driest areas of the world but also major economies and some of the world's wealthiest regions. The Western United States and the city of Cape Town are but a few recent examples.

The monitoring of SDG 6 on clean water and sanitation shows only moderate improvements in, for example, Africa, where millions of people on the continent still practise open defaecation (SDR 2021). Sanitation is a huge challenge and has been identified by the African Union as one of the main African challenges in the years to come.

Climate change mitigation may have adverse socio-economic impacts. One of the most severe effects is migration to urban centres and towards the north to reach the Mediterranean and across Europe. Thousands of people have died en route to a better life in the north, and mitigating the impacts of climate change remains one of the most significant global challenges facing us in the years to come.

The adaptation to potential changes in the availability of water supplies because of climate change is estimated to be 9–11 billion USD per year in 2030. However, the total adaptation costs for managing increased flood risk, maintaining water quality standards, and supporting instream economic and environmental uses will increase to a staggering 32–40 billion USD annually (UNFCC 2007).

Climate change impacts have vast consequences in the water sector, increasing health risks, environmental pollution, and substantial economic impacts. While most water professionals are working on abatement and impact-minimising activities, it still does not look like all challenges will be solved in the near future. Sound medium- and long-term planning and firm political commitment will be required to minimise the negative consequences that mankind may face with climate change impacts.

There are numerous global, regional, and national initiatives to address the technical and socio-economic challenges associated with the impacts of climate change on water. The Water and Society project, involving four countries in Africa, three from Asia, and one from Europe (www.WaSoProject.org), is an example and has inspired this special issue, which presents some water management challenges in Asia and Africa resulting from climate change. Selected studies on changes in water quantities and their socio-economic impacts, as well as examples of abatement activities and their effects, illustrate the challenges involved, their status, and solutions.

Guest Editors

H. Ratnaweera IWA
Department of Building and Environmental Technology, Norwegian University of Life Sciences, Norway
E-mail: harsha.ratnaweera@nmbu.no

T. Sætersdal IWA
Center for International Health, University of Bergen, Bergen, Norway
E-mail: Tore.Satersdal@uib.no

S. B. Weerakoon IWA
Department of Civil of Engineering, University of Peradeniya, Peradeniya, Sri Lanka
E-mail: sbweera@pdn.ac.lk

F. M. Mutua IWA
Department of Meteorology, University of Nairobi, Nairobi, Kenya
E-mail: fmutua@uonbi.ac.ke

REFERENCES

BBC 2021 *Advantages of Global Warming – Local and Global Effects of Climate Change – Higher Geography Revision – BBC Bitesize.* Available from: https://www.bbc.co.uk/bitesize/guides/zwjjjxs/revision/7.

Church, J. A. & White, N. J. 2011 Sea-level rise from the late 19th to the early 21st century. *Surveys in Geophysics* **32** (4–5), 585–602. https://doi.org/10.1007/S10712-011-9119-1.

ESCAP 2021 *Ready for the Dry Years Building Resilience to Drought in South-East Asia, Second Edition.* Available from: www.asean.org.

IFRC 2020 *Tackling the Humanitarian Impacts of the Climate Crisis Together World Disasters Report 2020.* Available from: https://media.ifrc.org/ifrc/world-disaster-report-2020.

IPCC 2014 *AR5 Synthesis Report: Climate Change 2014 – IPCC.* Available from: https://www.ipcc.ch/report/ar5/syr/.

Kastl, G., Sathasivan, A. & Fisher, I. 2016 A selection framework for NOM removal process for drinking water treatment. *Desalination and Water Treatment* **57** (17), 7679–7689. doi:10.1080/19443994.2015.1044476.

Kistemann, T., Claßen, T., Koch, C., Dangendorf, F., Fischeder, R., Gebel, J., Vacata, V. & Exner, M. 2002 Microbial load of drinking water reservoir tributaries during extreme rainfall and runoff. *Applied and Environmental Microbiology* **68** (5), 2188–2197. https://doi.org/10.1128/AEM.68.5.2188-2197.2002.

Kritzberg, E. S., Hasselquist, E. M., Škerlep, M., Löfgren, S., Olsson, O., Stadmark, J., Valinia, S., Hansson, L. & Laudon, H. 2020 Browning of freshwaters: consequences to ecosystem services, underlying drivers, and potential mitigation measures. *Ambio* **49**, 375–390. https://doi.org/10.1007/s13280-019-01227-5.

Mekonnen, M. M. & Hoekstra, A. Y. 2016 Sustainability: four billion people facing severe water scarcity. *Science Advances* **2** (2), e1500323. https://doi.org/10.1126/sciadv.1500323.

MGI 2020 *Climate Risk and Response: Physical Hazards and Socioeconomic Impacts.* Available from: www.mckinsey.com/mgi.

Pfahl, S., O'Gorman, P. A. & Fischer, E. M. 2017 Understanding the regional pattern of projected future changes in extreme precipitation. *Nature Climate Change* **7** (6), 423–427. https://doi.org/10.1038/nclimate3287.

Pokhrel, Y., Felfelani, F., Satoh, Y., Boulange, J., Burek, P., Gädeke, A., Gerten, D., Gosling, S. N., Grillakis, M., Gudmundsson, L., Hanasaki, N., Kim, H., Koutroulis, A., Liu, J., Papadimitriou, L., Schewe, J., Schmied, H. M., Stacke, T., Telteu, C., Thiery, W., Veldkamp, T., Zhao, F. & Wada, Y. 2021 Global terrestrial water storage and drought severity under climate change. *Nature Climate Change* **11** (3), 226–233. https://doi.org/10.1038/s41558-020-00972-w.

SDR 2021 *2020 Africa SDG Index and Dashboards Report – Sustainable Development Report.* Available from: https://www.sdgindex.org/reports/2020-africa-sdg-index-and-dashboards-report/.

UNDRR 2021 *GAR Special Report on Drought 2021 | UNDRR.* Available from: https://www.undrr.org/publication/gar-special-report-drought-2021.

UNFCC 2007 *Investment and Financial Flows to Address Climate Change.* United Nations Framework Convention on Climate Change, p. 273.

Valdivia-Garcia, M., Weir, P., Graham, D. W. & Werner, D. 2019 Predicted impact of climate change on trihalomethanes formation in drinking water treatment. *Scientific Reports* **9** (1). https://doi.org/10.1038/s41598-019-46238-0.

WHO 2010 *The Resilience of Water Supply and Sanitation in the Face of Climate Change Technical Report.* Public Health and Environment Water.

WMO 2021 *State of the Global Climate 2020 (WMO-No. 1264).* WMO, Geneva.

Water Management Addressing Societal and Climate Change Challenges doi: 10.2166/wcc.2021.351

Assessment on the economic damage of trading households due to urban inundation

B. V. Hung[a,*] and N. N. Diep[b]

[a] University of Science, Vietnam National University, Ho Chi Minh City, Vietnam
[b] University of Labour and Social Affairs, Secondary Branch, Ho Chi Minh City, Vietnam
*Corresponding author. E-mail: bvhung@hcmus.edu.vn

BVH, 0000-0003-4034-7888

ABSTRACT

Floods impact most of the subjects living and producing in Ho Chi Minh City (HCMC). Trading households are one of the important economic sectors in the economic development of HCMC. This sector contributes greatly to development from GDP to creating jobs and attracting a large workforce cooperating with them such as goods traffic and business services. Because the trading households are of small business scale and very diverse, they are often scattered in residential areas, along roads. They are also adversely affected and suffer economic losses due to inundation, like citizens. The determination of damage level to all objects including trading households is one of the flood management activities and the basis for infrastructure development projects as well as social support policies of the city. Through a direct survey of more than 3,419 trading households and applying the economic loss curve method, this research has established a correlative function for economic losses due to inundation with flood depth and duration. With the correlative function of flood damage for trading households, the district and city's agencies will have a new tool to assess and analyze the damage level of trading households selling in the area.

Key words: correlative function, damage function, flood damage, trading business, urban inundation

HIGHLIGHTS

- It has found the local damage types due to urban flood caused.
- It has established the flood damage formula and map for HCMC.
- The flood damage formula included two flood factors (depth and duration) with their weights.
- It is the first research on flood damage for the whole of HCMC with the object being the selling householder.
- The research determinated the flood depth level able cause damage for HCMC (>15 cm).

INTRODUCTION

Ho Chi Minh City (HCMC), one of the largest cities in Vietnam, currently faces many shortcomings in the process of socio-economic development, and one of these manifestations is that many economic indicators are out of reach compared with the city's development plans due to problems such as floods, traffic jams, and illness. The inundation problem has been going on for many years in HCMC for various reasons such as heavy rainfall, high tides, poor drainage system in the old residential area, and ground subsidence due to groundwater extraction.

Urban inundation is one of the problems that seriously affects the economy of big cities in the world as well as in Vietnam. Everyone, from enterprises to trading households and households including the poor or wealthy are also affected by inundation. Citizens are facing the impacts of inundation on the economic sector, and the government is facing the financial deficits of adapting to the flood situation. Trading households are an important economic component due to their great contributions to the overall economic development of the area and to job creation for people. The trading households are of diverse scale and type of business, so they often interweave and share the same infrastructure with residential areas. Therefore, trading households are not only affected by the same damage as households (construction, electricity, drainage system) but also suffer economic losses in production, business, and labor problems (stopping production and business because workers are late or take leave due to floods, etc.). This problem leads to a trading household having a much larger loss level and being more vulnerable than others.

In order to assess damage caused by inundation, it is first necessary to clarify the risks due to an inundation situation. Risk analysis is seen as an activity of flood management, and is also used in analyzing the different types of damage that urban inundation causes to the objects affected. Therefore, this analysis is important when assessing flood disaster mitigation options and carrying out the economic optimization of possible measures (Apel *et al.* 2004). One way to represent regional flood damage is often quantified by using a flood damage assessment.

Assessment of economic losses due to urban inundation has been carried out by previous studies using many different approaches; with loss assessments (Merz *et al.* 2011) based on a future risk map approach and a financial appraisal approach for insurance and compensation for present damage. Economists (Olesen *et al.* 2017) have identified two main ways to assess the damage cost of a natural hazard; to assess economic losses and financial losses. The costs incurred for flood restoration and improvements, the substantial costs that remain to repaired damaged infrastructure and urban services in urban areas are considered for determining the economic damage level (The Center for Neighborhood Technology 2014). The calculation of damage caused by inundation is also based on mathematical modeling to determine the inundation level (usually flood depth) with the area of each different land-use class affected (Bao & Cong 2016).

The importance in assessing general disaster damage and special urban inundation is determined by the assessment of damage groups. The damage groups are often classified into direct and indirect damage, tangible and intangible damage (Merz *et al.* 2011). For trading households, the relative damage groups can be identified as:

- Damage is determined to be direct (Handmer *et al.* 2002) when costs to repair the damages occur as a result of direct contact with water. In the city, elements such as walls, floors, yards and the assets of houses are damaged along with goods, materials of stores, factories, facilities, etc. (Bao & Cong 2016).
- Damage is determined to be indirect (Price & Vojinovic 2008) when the costs incurred are as a consequence of inundation impacts in flood duration or afterwards. They involve a slowing of or stopping production and business, causing financial difficulties, as well as adversely affecting the mental health of workers, taking extra time to transport goods, and transactions and the demand production market lost due to traffic jams or seriously damaged roads (Zhu *et al.* 2018).
- Damage is also considered to be tangible (Handmer *et al.* 2002) when it can be objectively quantified, i.e., damage can be calculated as a direct monetary value such as costs of repairmen and overcoming the consequences of inundation.
- Damage is considered to be intangible when it is related to the health and well-being of employees, loss of investment opportunities, profits, etc. Intangible damage sometimes includes that to human life and loss of cultural properties (Green *et al.* 2011).

The damages caused by flood are of small value in cities compared with the regions. However, floods happen regularly in coastal cities like HCMC so the total damage is significant compared with citizens' income (Green *et al.* 2011; Bao & Cong 2016). The damage types are shown in Table 1.

Table 1 | Classification of damages due to flood causes for a trading household

Type	Directness	Indirectness
Tangible	Construction element damages	Traffic jams
	Traffic facilities	Business interrupted
	Infrastructure system (drainage)	Support to workers affected
	Properties, materials, goods	Living interrupted
	Lift up, arranging, cleaning activities	Sale revenue declines
Intangible	Worker healthiness	Tense nerves and worry
	Disease	Daily living/working schedule interrupted
	Losses at business, production places	Workers leave, are sick
	Regional environmental damages	Land value decreased
	Inconvenience for workers	Mental health declines

Source: Handmer *et al.* (2002); Hammond *et al.* (2015).

When describing the inundation situation/size, many researchers have often applied flood factors such as velocity flow, flood depth, flood duration, flood area, etc. Based on the available data or ability to collect data and information on the study region, research can apply one or a few factors (Price & Vojinovic 2008; Bao & Cong 2016). The flood depth is applied in most domestic and international studies, because it is usually easily determined at hydrological stations.

According to Olesen *et al.* (2017), in most of studies in the world, it quantifies the direct/tangible as well as indirect/intangible damage according to three general methods, which are:

- The assessment of flood damage according to the insurance data paid for previous flood events (Olesen *et al.* 2017).
- The common cost method (applying the types of land use as cost per area unit) (Bao & Cong 2016).
- The curve method between flood factor (depth) and damages (a depth-damage curve is a flood-damage function) (Price & Vojinovic 2008).

The depth-damage curve can determine the direct (tangible) or indirect (intangible) damages caused by the flood (the representative depth factor) (Albano *et al.* 2015). Both direct (tangible) and indirect (intangible) damages to subjects performed by depth-damage curves can be identified through the integration of survey data or the relative experience. To construct an accurate depth-damage curve for an area, a great deal of data on concrete flood factors and the corresponding damage are required (Albano *et al.* 2015).

To determine the extent of damage caused by flooding to a subject, the depth-damage curve method – the flood-damage function – is widely used in domestic and foreign studies. Urban flood depth is considered to be an important factor when considering physical damage (Handmer *et al.* 2002). The formula of damage correlation (D) with flood depth (h) commonly uses the following form (Bao & Cong 2016):

$$D = \sum_{i=1}^{n} F_i \times f(h_i) \tag{1}$$

where D is total damage to the research area, n is number of cells of the research area, F_i is area of cell number i-th and $f(h_i)$ is damage value of cell number i-th correlated with flood depth (h_i) determined by damage function.

Besides the flood's depth factor, it also has many other factors, in which duration is the factor that increases the extent of damage (Price & Vojinovic 2008). With the same level of flood depth, the longer the duration the greater the damage. The formula to calculate the tangible damages applies the two main flood factors which are flood depth and duration. This can make the assessment of damage become more reasonable and exacting (Price & Vojinovic 2008).

$$D = A_0 + A_1 H + A_2 T \tag{2}$$

where T is duration, and A_0, A_1, A_2 are the parameters determined for the different land use.

In Viet Nam, according to the guidelines of the Ministry of Construction on defining urban inundation, the urban inundation factors are defined by the depth and duration (MC 2003). Each inundation factor is determined by certain intervals representing the inundation's levels. The location is defined as an inundation situation when the flood factors satisfy the following conditions with flood depth of around 30 cm and flood duration longer than 30 min. Therefore, the local administrations usually integrate all flood factors including depth and duration, which are investigated by the actual inundation event. Formula (2) is according to the present management situation in Viet Nam.

It is difficult to identify general flood duration in flooding simulation. Thus, in the flooding model, the result is usually the flood depth (Bao & Cong 2016). Formula (1) is applied in the most relative research in the world as well as Viet Nam and particularly HCMC. In fact, a flood's duration is a reason for causing the extent of damage to increase. Formula (1) should represent this increasing extent of damage. Hence, Formula (1) has changed when taking into account the inundation duration factor in the damage assessment. The inundation depth is a conversion inundation depth ($H_{cover}=f(h,T)$). Therefore, Formula (1) has a similar consequent form to Formula (2).

Therefore, in the research scope as well as the journal's content, we intend to assess all direct and indirect/tangible and intangible damage levels due to urban flood for a trading household living and selling in HCMC. The flood-damage curve method with the two inundation factors of depth and duration are applied to establish the damage correlative formula in the study.

METHODOLOGY

To determine the damage for the analysis of a correlation between the damage and the flooding factors (flood depth and duration) and establish a correlation function of damage caused by urban inundation, we implemented a survey of economic damage caused by flooding for trading households selling in 22 districts of HCMC (see Figure 1). The sequence of research is done according to the steps below.

The size of survey and number of objects surveyed/interviewed

The surveying directly targets objects affected by floods and is implemented in 22/24 districts of HCMC. The number of trading households is determined by Yamane's formula:

$$n = \frac{N}{1 + N * e^{x}} \tag{3}$$

where n is number of samples, N is the total objects, and e is standard error (5%).

The number of HCMC's population for 22/24 districts is about 8,515,084/8,993,082 (approximately 2,128,771 households) (http://www.dubaonhanluchcmc.gov.vn/tin-tuc/8012.bao-cao-thi-truong-lao-dong-thanh-pho-ho-chi-minh-nam-2019—du-bao-nhu-cau-nhan-luc-nam-2020.html). The trading household number is about 64.89% (about 1,381,359). The number of samples is 400. However, the important condition of investigation is those trading households affected or damaged by the inundation.

It is not possible to determine exactly the total length of road flooded in HCMC. Therefore, in the study, we applied approximatively the number of 232.5 km in 2016–2018 and 109.4 km in 2019 from administrative districts of HCMC. Thus, the number of examples is 2,325 for the period 2016–2018 and 1,094 for 2019 (the study started in November 2019) with 100 m taken as one example.

The damage caused by urban inundation is determined as follows:

$$\textit{Total damage} = \textit{The direct damage} + \textit{The indirect damage}\ (\text{unit:}10^{6}\text{VND}) \tag{4}$$

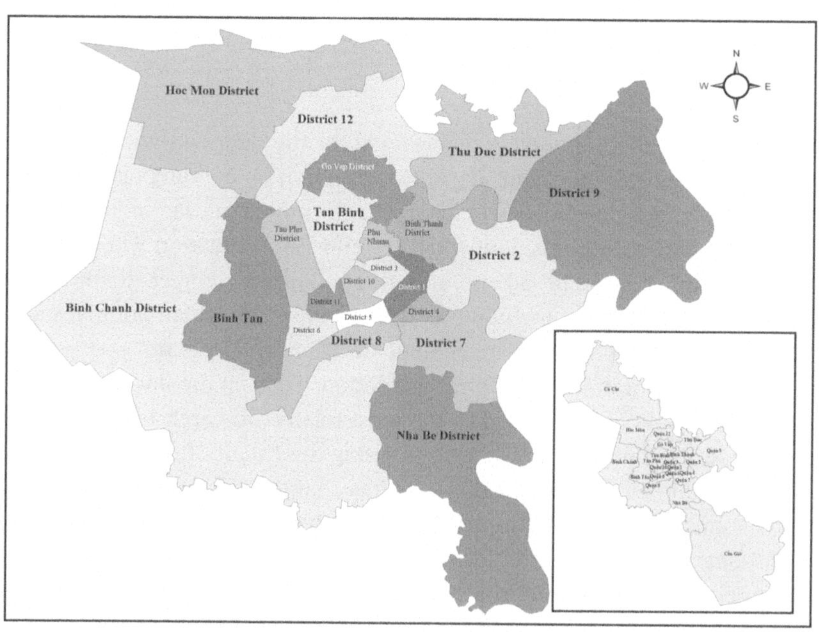

Figure 1 | The districts of HCMC where the survey was implemented.

To determine the flood-damage correlation function with the inundation depth (cm) and duration (min) factors, the study uses the regression method to establish and evaluate the degree of correlation (R^2). The correlation Formula (1) is applied to the research with the analysis of the questionnaire and determine the damage correlation function following the steps in Figure 2.

Figure 2 shows that Formula (2) of damage correlative function will be established after Formula (1) is defined. Based on the correlation formula between inundation depth (h) and duration (T), we will replace the inundation duration of Formula (1) by the duration function by inundation depth. Thus, the damage correlative function under Formula (2) is the function by the inundation depth. However, this inundation depth is now the conversion depth since the damage correlation function has taken into account the inundation duration factor representing indirect damage/intangible damage.

Therefore, the damage correlative functions following Formulas (1) and (2) are applied in the study as (unit: 10^6 VND):

The function by Formula (1): $D = A_o + A_h h(cm) + A_t T(min)$ (5)

The function by Formula (2): $D = B_o \operatorname{Ln}(H_{conver.}) + B$ and $H_{conver.} = (hx(a\operatorname{Ln}(h) + b)^{1/2}$ (6)

where D is total damage, h is flood depth (cm), T is duration (min), A_O, A_h, A_t, B_O, B, a, and b are the parameters of each component in Formulas (5) and (6).

RESULTS AND DISCUSSION

The survey results

After the survey activity was implemented in 22/24 districts of HCMC, the number of questionnaire papers completed and collected is summarized in Table 2. The numbers of questionnaire papers for each district are very different due to the

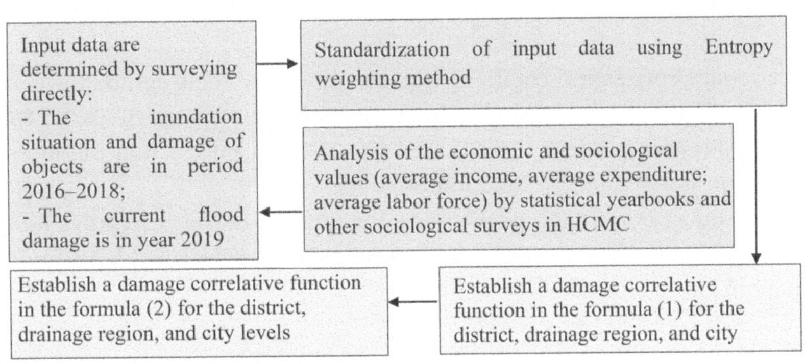

Figure 2 | The diagram of approach to the formulating flood damage.

Table 2 | Summary of questionnaire papers to survey the trading households in the districts

Name of district	2016–2018	2019	Name of district	2016–2018	2019
Dist. 2	17	36	Dist. 6	354	111
Dist. 4	27	42	Dist. 9	170	84
Thu Duc Dist.	32	71	Dist. 10	0	0
Dist. 7	66	30	Dist. 11	33	42
Dist. 8	44	65	Bình Thanh Dist.	88	50
Nha Be Dist.	45	12	Tan Binh Dist.	65	41
Dist. 12	163	65	Tan Phu Distr.	306	72
Go Vap Dist.	42	27	Phu Nhuan Dist.	46	49
Dist. 1	9	7	Binh Tan Dist.	302	45
Dist. 3	2	1	Hoc Mon Dist.	123	104
Dist. 5	11	13	Binh Chanh Dist.	357	86

differences of inundation conditions (between central and suburban, new and older districts). This is reasonable as it is considered that the infrastructure (drainage culvert system) is more complete in central districts compared with suburban ones.

Based on the information from the economic and social sectors (average income, average expenditure, average labor force) by statistical yearbooks and other sociological surveys in HCMC, the inundation damage of trading households affected is summarized in Table 3.

Based on the column 'Depth' of Table 3, the inundation damages occur and are valuable when the depth is about 15–45 cm and the duration is over about 30 min. The range of inundation depth identified from surveying is according to the urban inundation condition of Viet Nam Ministry of Construction (MC 2003).

The statistical analysis

The evaluation of the survey results on statistical significance as well as the square correlation level are based on determining the *P*-value and R^2 values (see Table 4). From there, it will be determined whether the survey results are eligible to be analyzed and set up a correlation regression function.

To determine the *P*-values of all survey data series we evaluated the samples of each data series as a damage series, inundation depth series, and inundation duration series. Their *P*-values are very small and lower than 0.05 (see bottom three rows of Table 4). This means there is a statistical significance.

Table 4 also defines the *P*-values of a combination of data series (damage, depth, and duration) in two periods (2016–2018 and 2019). They are very small. We also conclude that the survey data series has a statistical significance.

Therefore, all survey data series (inundation damage, depth, and duration) have the same trend (depth and duration coefficients are positive).

The square correlation coefficient R^2 between the damage level and the flooding factors including flood depth h (cm) and duration (min) for HCMC shows that their correlation is close (see Table 5). The square correlative coefficients are summarized in Table 5.

Table 5 also shows that the square correlation coefficients, both univariate (between damage and each inundation factor) and multivariate (between a damage with both inundation factors), are in the range of 50–80%. The square correlation coefficient between the inundation depth and duration is also close (50–80%). It means that the inundation depth and duration have a relationship or the duration is a significant function of depth.

With the number of the above square correlation coefficients, the formula of the correlative functions between the inundation damage and each flooding factor (depth/duration) as well as both flooding factors (depth and duration) will be nonlinear (the natural logarithmic function).

Determine the coefficient sets for damage correlative functions

Based on the results of direct and indirect damage obtained and the statistical analysis method, the inundation damage function applied to identify the damage of trading households affected in HCMC, following Formulas (5) and (6), has the coefficient sets as shown in Table 6.

Therefore, the correlative functions between inundation damage and two factors (depth and duration), as well as between damage and conversion depth, are defined below:

The damage function (Formula (5)):

$$D_{H,T} = -1.762 + 0.067\,h(\text{cm}) + 0.036\,T(\text{min})$$

The damage function (Formula (6)) with depth 15 cm–30 cm:

$$D_{H\text{conver.}} = 3.044\ \text{Ln}(H_{\text{convert}}) - 9.05$$

The damage function (Formula (6)) with depth 30 cm–45 cm:

$$D_{H\text{conver.}} = 5.494\ \text{Ln}(H_{\text{convert}}) - 18.246$$

Table 3 | The inundation damages of trading households in the period 2016–2019

Locations		Flood factors			Analysis properties			Tangible damages							Intangible damages						
		Depth	Duration	number	House	Assets	Goods	House	Yard	Under constr.	Assets	Vehicles	Equipments	Goods	Clearning time	Traffic jam time	Delay	Not sell	Sick cost	Remedial cost	Others
No	Number	cm	minute	time	10^9VND	10^9VND	10^9VND	10^6VND	10^6VND	10^6VND	10^6VND	10^6VND	10^6VND	10^6VND	date	date	date	date	10^6VND	10^6VND	10^6VND
Ho Chi Minh																					
1	Q12-01	43	98	6	1.0	0.1	0.0	7.0	3.0	1.0		5.0	25.0		2.0	2.0			2.0		5.0
1	Q12-01	28	52	13	1.0	0.1	0.0	2.5	2.0	1.0	1.0	1.0	7.0	2.5	2.0	1.0	2.1		1.0	5.0	5.0
1	Q12-01	28	52	13	2.0	0.1	0.0	2.5	2.0	1.0	1.0	1.0	7.0	2.5	2.0	1.0	2.1		1.0		5.0
1	Q12-01	23	30	13	2.0	0.1	0.0	2.5	2.0			1.0	5.0		3.0	5.0				5.0	
1	Q12-01	23	30	13	4.0	0.1	0.1	2.5		1.0	1.0			2.5	3.0	1.0	2.0	1.0	1.0		5.0
2	Q12-02	25	53	11	2.0	0.1	0.0	7.0	2.0	1.0	5.0	5.0	10.0	2.5	2.0		2.2	1.0	2.0	10.0	5.0
3	Q12-03	26	44	13	2.0	0.1	0.0	2.5	2.0	1.0	1.0	1.0	7.0	2.5	2.0	1.0	2.1	1.0	1.0	5.0	5.0
3	Q12-03	29	51	12	1.0	0.1	0.0	2.5	2.0	1.0	1.0	1.0	7.0	2.5	3.0	2.0	2.1	1.0	2.0	5.0	5.0
3	Q12-03	26	37	13	4.0	0.1	0.0	2.5	2.0	5.0	1.0	1.0	7.0	2.5	2.0	1.0	2.1	2.0	2.0	5.0	5.0
3	Q12-03	40	120	6	2.0	0.1	0.0	7.0	3.0	1.0	1.0	5.0	12.0	2.5	3.0	5.0	5.1	2.0		5.0	5.0
3	Q12-03	29	51	12	4.0	0.1	0.0	2.5	2.0	1.0	1.0	5.0	7.0	7.5	2.0	8.0	0.1	1.0		5.0	5.0
3	Q12-03	40	120	6	4.0	0.1	0.1	2.5	3.0	1.0	1.0	5.0	12.0	2.5	3.0	4.0	5.1	2.0	2.0	5.0	10.0

Table 4 | The synthesizing results of statistical analysis for the period 2016–2019

Summary output				**ANOVA**						
(2016–2018) *Regression statistics*		**(2019) *Regression statistics***			**(2016–2018)**			**(2019)**		
					df	***F***	***P-value***	***df***	***F***	***P-value***
Multiple R	0.803737	Multiple R	0.80730	Regression	2	2,098	0	2	982	3.2×10^{-241}
R square	0.645992	R square	0.65173	Residual	2,299			1,050		
Adjusted R square	0.645684	Adjusted R square	0.65107	Total	2,301			1,052		
Standard error	1.159925	Standard error	1.10489							
Observations	2,302	Observations	1,053							
	Coefficients	***Standard error***	***t Stat***	***P-value***						
Intercept	**−2.44948**	0.17952	−13.6206	5.137×10^{-40}						
Depth	**0.10917**	0.00928	11.7186	2.124×10^{-29}						
Duration	**0.05925**	0.00226	26.6527	1.513×10^{-73}						

Table 5 | The square correlation coefficients (R^2) between an inundation damage and factors and between two inundation factors

Object: Trading household		**The square correlative coefficient (R^2)**					
		Period 2016–2018		**Year 2019**		**Duration/Depth**	
No.	**Location**	**Depth**	**Duration**	**Depth**	**Duration**	**Depth**	**Duration**
1	Ho Chi Minh City	51%	54%	52%	53%	64%	51%

Table 6 | The coefficient sets of inundation damage functions following Formulas (5) and (6)

Trading household		**Flood damage function by:**			**15–30 cm**		**30–45 cm**		**15–45 cm**	
		$D_{H,T}$			$D_{Hconvert}$		$D_{Hconvert}$		$D_{Hconvert}$	
No.	**Name**	***Ao***	***Ah***	***At***	***Bo***	***B***	***Bo***	***B***	***Bo***	***B***
1	Ho Chi Minh	−1.76	0.067	0.036	3.04	−9.05	5.49	−18.24	4.043	−12.42

With the conversion depth:

$$H_{\text{convert}} = (h \times (97.3\ \text{Ln}(h) - 268.4))^{1/2}$$

With the condition ranges of inundation depth (15 cm – 30 and 30 cm – 45 cm), the damage-depth curves represent the correlative functions (Formula (6)) between the damages to trading households caused by inundation and the conversion depths, which is determined by survey flooding depth and duration. These damage-depth curves are shown in Figures 2 and 3.

Figures 3 and 4 also show that although the square correlation coefficients R^2 of the damage and conversion depth are still very close (close to 1 unit), when the range of depth is wider, the parameter B_o and B of the damage function (Formula (6)) also varies. However, the absolute rate 'slope value' of the damage-depth curves decreases gradually corresponding to the depth intervals increasing. This proves the reasonableness of the damage-conversion depth curve (including the duration factor into function) when the inundation depth increases cause the value of damage to increases with velocity to be reduced and approach the property value of the survey object.

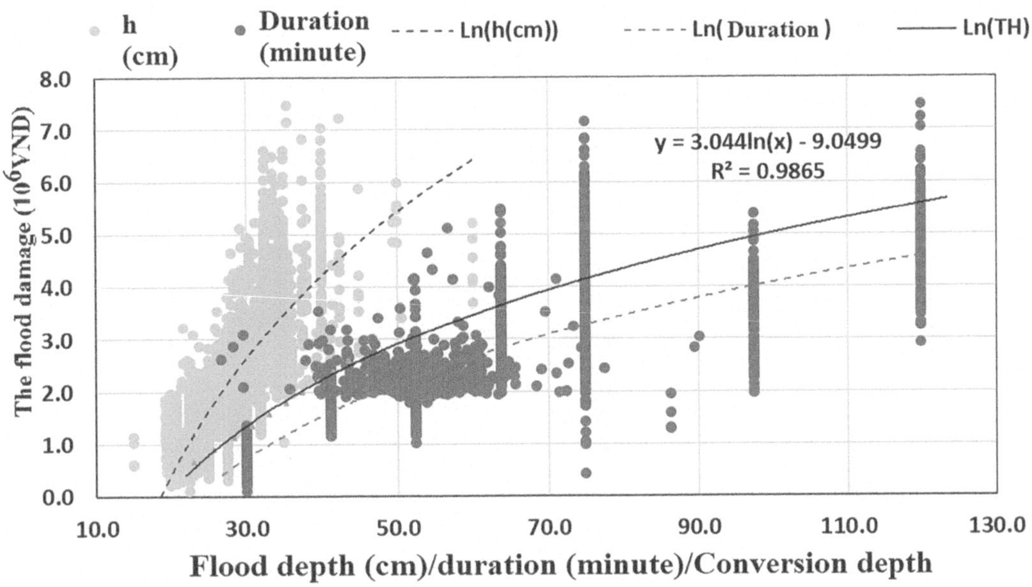

Figure 3 | The flood-damage curve corresponding to the surveyed depth from 15 to 30 cm.

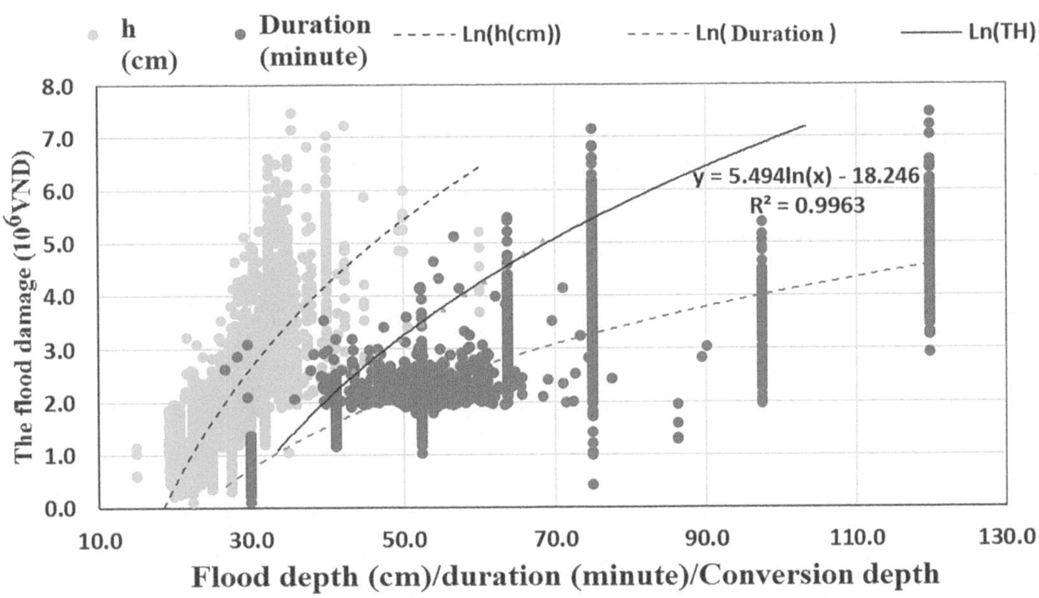

Figure 4 | The flood-damage curve corresponding to the surveyed depth from 30 to 45 cm.

To calculate the ratio of (B_o/B) of the damage functions corresponding to the depth intervals shown in Figures 3 and 4, the ratio values are 0.343 and 0.306, respectively. With the inundation depth range varying from medium to heavy flooding intervals (15–45 cm), the 'slope' of this damage curve is the average number (0.331) of the two 'slopes' of medium and heavy flood depths.

Thus, in the study, when applying the flood damage formula in Formula (6) to the simulative model, we can apply the average parameter of two damage levels (medium and heavy) corresponding to the flood depth around 15–45 cm (covering all damage of medium and heavy level). The flood-damage correlative function and figure for trading households selling in the HCMC is shown in Figure 5.

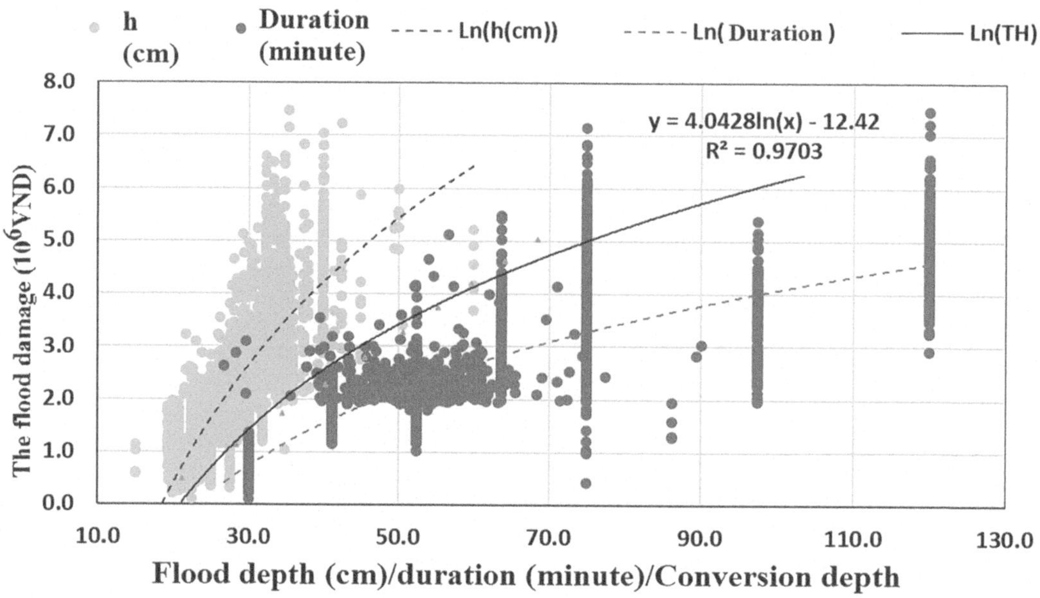

Figure 5 | The flood-damage curve corresponding to the surveyed depth from 15 to 45 cm.

Based on the investigation of the flood situation in 2019, we used the flood-damage function to assess the flood-damage value of trading households affected in District No. 1 of HCMC as an example to establish a flood-damage distributive map (see Figure 6).

Finally, the flood-damage functions applied to assess the damages caused by the urban inundation in HCMC such as Formula (5) is used in the urban management (administration) and Formula (6) uses the results (depth) of the simulation inundation model to identify the damages to trading households in HCMC:

Formula (5): $D_{H,T} = -1.762 + 0.067\,h(\text{cm}) + 0.036\ \text{T (minute)}$

Formula (6): $D_{H\text{conver.}} = 4.0428\ \text{Ln}(H_{\text{convert}}) - 12.42$

With $H_{\text{convert}} = (h \times (97.3\ \text{Ln}(h) - 268.4))^{1/2}$

Based on the calculation method and the results of the damage correlative function with the flooding factors, some assessments are as follows:

- Using a regression correlative function between the flood damage investigated and the two flooding factors (depth, duration) determined both correlation and damage function. Although the flood-damage functions are set up in two different formulas, in fact, the flood-damage Formula (5) is a consequence of Formula (4) when the correlation function is determined between two flooding factors, namely, inundation duration and depth.
- The correlation coefficient value (R^2) between the flood damage of the trading household affected/damaged by urban inundation and the flooding factors have determined from around 50% to 65% to ensure that their correlations are close. This means that the survey data series of flood damage, flood depth, and duration have the same trend and their correlative functions are in curvilinear regression form.

CONCLUSIONS

In the past ten years, HCMC has taken many actions to reduce urban inundation. However, the local inundation situation still occurs. With it comes more heavy rain, and the tidal peak increases continuously leading to the inundation situation becoming more serious and having significant impacts on the peoples' living and selling activities.

Through a field survey of the flooding damage (direct/indirect; tangible/intangible) caused by inundation affecting a trading household object, the initial research has established a damage correlative function due to inundation for HCMC. The

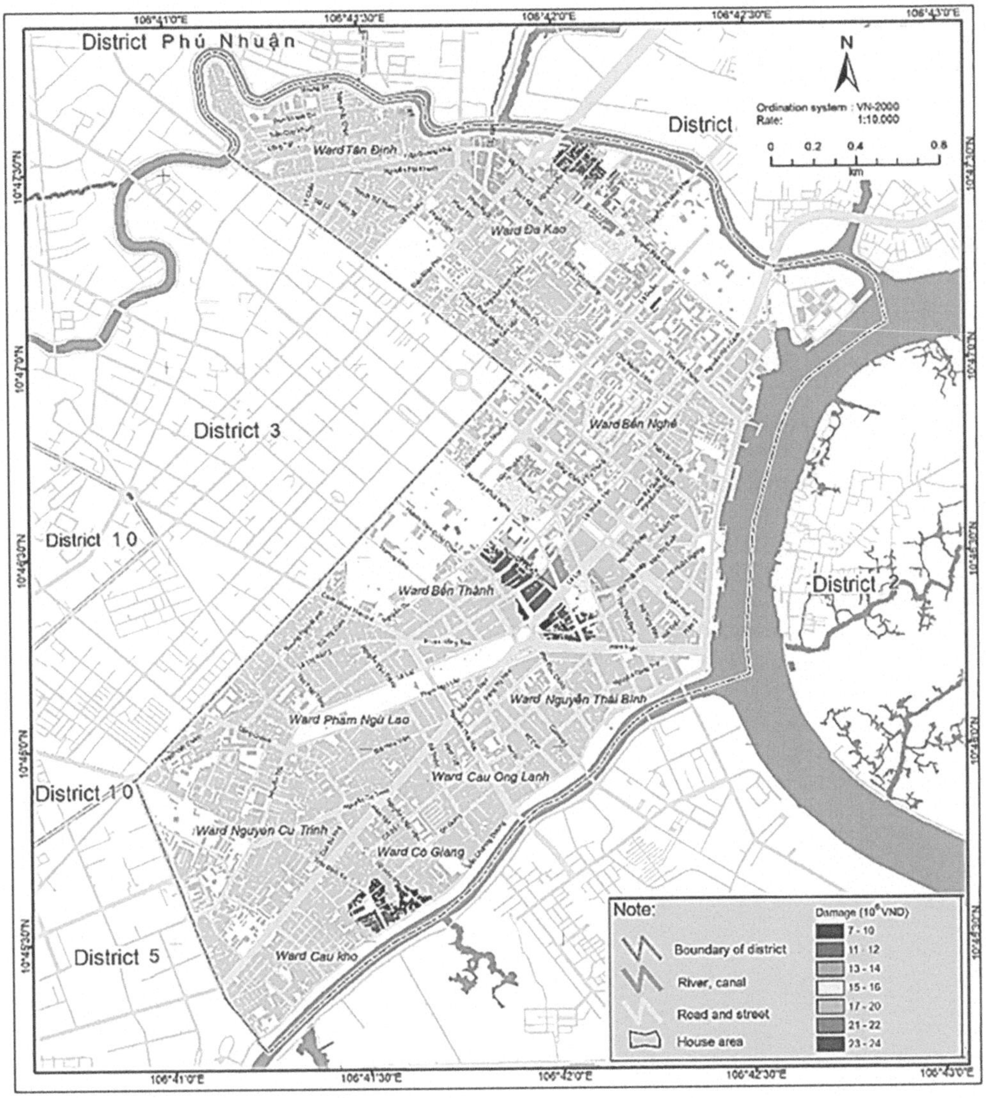

Figure 6 | The flood-damage distribution of selling households at District No. 1 of HCMC (period 2018–2019).

analysis results show that the damage correlative functions following Formulas (1) and (2) meet the determination of the economic loss of a trading household object selling in a city adversely affected by inundation.

Due to the weather in HCMC being complicated and becoming more extreme, the flooding situation continues to evolve unpredictably and causes a lot of damage to the subjects in both extent and type.

The study of flood damage assessment stopped establishing the form of correlative function between the damage with the flooding factors and initially determined the set of coefficients for the period from 2016 to 2019. To make the formula more stable, it is necessary to continue the survey for the following years (possibly with a narrower scale).

ACKNOWLEDGEMENTS

The paper has been carried out with results from the research 'Survey and assessment of damage caused by an inundation to socio-economic; mapping a current flood damage for the prevention of flooding, urban planning in Ho Chi Minh City'.

DATA AVAILABILITY STATEMENT

Data cannot be made publicly available; readers should contact the corresponding author for details.

REFERENCES

Albano, R., Mancusi, L., Sole, A. & Adamowski, J. 2015 Collaborative strategies for sustainable EU flood risk management: FOSS and geospatial tools – challenges and opportunities for operative risk analysis. *ISPRS International Journal of Geo-Information* **4**, 2704–2727.

Apel, H., Thieken, A., Merz, B. & Blöschl, G. 2004 Flood risk assessment and associated uncertainty. *Natural Hazards and Earth System Sciences* **4**, 295–308.

Bao, L. X. & Cong, M. V. 2016 Economic risk assessment due to flooding, applying to the flood control project phase 1 in HCMC. *Journal of Science, Technology of Irrigation and Environment* **55**, 55–72.

Green, C., Viavattene, C. & Thompson, P. 2011 *Guidance for Assessing Flood Losses. CONHAZ Report WP06.* Middlesex University, UK.

Hammond, M. J., Chen, A. S., Djordjević, S., Butler, D. & Mark, O. 2015 Urban flood impact assessment: A state-of-the-art review. Urban Water Journal **12** (1), 14–29, DOI: 10.1080/1573062X.2013.857421.

Handmer, J., Reed, C. & Percovich, O. 2002 *Disaster Loss Assessment Guidelines.* Department of Emergency Services, State of Queensland, Australia, p. 90.

Merz, B., Kreibich, H. & Lall, U. 2011 Multi-variate flood damage assessment: a tree-based data-mining approach. *Natural Hazards and Earth System Sciences* **13** (1), 53–64.

Ministry of Construction (MC) 2003 *Official Letter No. 338/BXD-KTQH About Guideline on the Defining Urbun Inundation.* Viet Nam Government.

Olesen, L., Löwe, R. & Arnbjerg-Nielsen, K. 2017 *Flood Damage Assessment – Literature Review and Recommended Procedure.* Cooperative Research Centre for Water Sensitive Cities, Clayton, Victoria, Australia. Available from: https://watersensitivecities.org.au/content/flood-damage-assessment-literature-review-recommended-procedure/

Price, R. K. & Vojinovic, Z. 2008 Urban flood disaster management. *Urban Water Journal* **5**, 259–276.

The Center for Neighborhood Technology 2014 'The Prevalence and Cost of Urban Flooding: A Case Study of Cook County, IL' (32pp).

Zhu, J., Dai, Q., Deng, Y., Zhang, A., Zhang, Y. & Zhang, S. 2018 Indirect damage of urban flooding: investigation of flood-induced traffic congestion using dynamic modeling. *Water* **10** (5), 622.

First received 23 December 2020; accepted in revised form 11 August 2021. Available online 21 September 2021

Water Management Addressing Societal and Climate Change Challenges doi: 10.2166/wcc.2021.122

Modernization of the left bank irrigation system of the Kaudulla reservoir of Sri Lanka

W. D. M. Wijenayaka [a,*], S. B. Weerakoon [a] and Francis Mutua [b]

[a] Department of Civil Engineering, University of Peradeniya, Kandy, Sri Lanka
[b] College of Biological & Physical Science, University of Nairobi, Nairobi, Kenya
*Corresponding author. E-mail: cengmahinda@gmail.com

WDMW, 0000-0001-8832-4066; SBW, 0000-0002-3671-6241

ABSTRACT

Irrigation water management in the paddy cultivation area under the left bank canal of the Kaudulla reservoir (Kaudulla tank) in the North Central Province of Sri Lanka has become a serious issue due to limited water availability and inefficient water distribution infrastructure at present. Insufficient storage capacities of the village tanks in the cultivation area, low rainfall during some months and regulated inflow from the transbasin diversion to Kaudulla tank have had a significant impact on yield in the two cultivation seasons of the year. In this article, modernization of irrigation infrastructure in the command area was investigated for effective utilization of limited available water. The results of the calibrated and validated Hydrologic Engineering Centre - Hydrologic Modeling System (HEC-HMS) model and Crop Water and Irrigation Requirements Program (CROPWAT) model were used with Water Evaluation and Planning Model (WEAP) to evaluate the water balance and demand to identify the best investment for improving irrigation water supply to maximize the return. Economic analysis was carried out using the net present values for different modernization options. Accordingly, the construction of a new canal system and augmentation of the capacities of village tanks from the present total capacity of 3.8–20 MCM was found to be the most appropriate option. This intervention will increase the income from yield by 205 million with a payback period of 12 years in the Yala season.

Key words: CROPWAT, irrigation water management, Kaudulla reservoir, paddy cultivation, WEAP

HIGHLIGHTS

- Insufficient storage capacities of minor tanks, low rainfall during some months in the tank catchments and regulated diversion to the Kaudulla tank have had a significant impact on the cultivation area under Kaudulla tank situated in the North Central Province of Sri Lanka.
- A calibrated and validated HEC-HMS model and the CROPWAT model were used with WEAP to evaluate the water balance and demand to identify the best investment for improving the irrigation system based on the economic return.

INTRODUCTION

The Kaudulla reservoir (the Kaudulla tank) and the areas irrigable from the tank fall under Mahaweli system ‘D’ and are situated in Medirigiriya District Secretariat Division of the North Central Province of Sri Lanka (Figure 1). Drought is frequently experienced in the region; for example, 1,110 people from 350 families in Medirigiriya were affected by drought in the latter part of 2017 (Jayarathne 2016). Water scarcity for irrigated paddy cultivation in the area has led to poor cropping intensity, especially in the Yala (April–August) season, which is one of the two seasons in which paddy is cultivated every year, the other being Maha (November–March). An irrigation water management study implemented to investigate the efficiency under normal operating conditions at Kaudulla indicates that water management is comparatively efficient during Yala even with the limited water resource, but that rainfall over the demand area could be used more effectively as a substitute for irrigation water during the Maha season (Holmes *et al.* 1978). A case study carried out centering on the Wadigewewa and Nikahena small tanks in the irrigation area strongly recommends an increase of the tank capacities to store the ample rainwater received in the rainy season (Rekha 2014).

The existing cultivation areas amounting to 7,160 ha have the potential to receive water from the left bank (LB) canal of the Kaudulla reservoir (the Kaudulla tank). However, only 3,790 ha are supplied directly by the left bank canal at present, and the rest is by 35 rainfed small village tanks. Limited storage capacities of the tanks, low rainfall in the dry season and regulated inflow

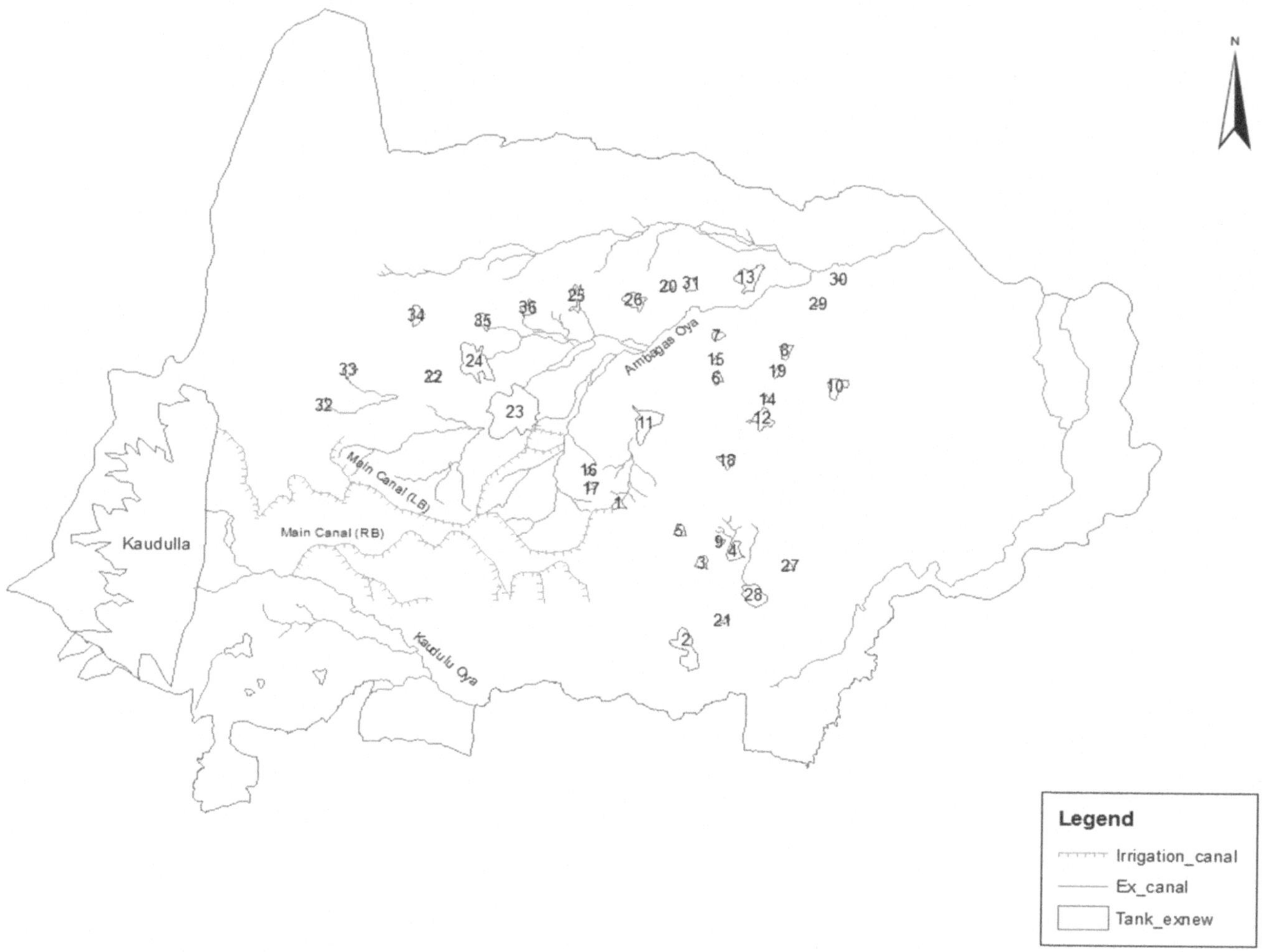

Figure 1 | Existing irrigation system used for the WEAP model.

to the Kaudulla tank have caused a significant temporal variation in paddy yield in the Kaudulla LB command area. It is against this background that this study is being carried out to propose appropriate modernization of the existing irrigation system to increase the yield derived from the LB command area. The study develops a model for water management in the left bank area to select the best intervention to maximize the return from investment while producing satisfactory yield under limited water availability, to support the farming families. The HEC-HMS model was applied to calculate the catchment yields of 35 village tanks. The results of the crop water requirement generated from the CROPWAT model were used with WEAP to evaluate the water balance and demand area coverage to identify the best investment for improving irrigation water supply to maximize the return. The best intervention among a few feasible structural interventions was identified based on the economic return.

METHODOLOGY

A calibrated and validated rainfall–runoff model generated using HEC-HMS was applied to calculate the catchment yields of 35 village tanks under the left bank canal of the Kaudulla tank (Figures 1 and 2; Table 1). The results of the crop water requirement generated from the CROPWAT model were used with WEAP to evaluate the water balance and demand area coverage.

An analysis was carried out using the CROPWAT and WEAP model applications to identify the best investment for improving irrigation water supply to maximize the return.

Method of analysis

LB canal irrigation areas are supplied with water by different methods. The cultivation area at the side of the first section of the LB canal is supplied with water directly from the canal through the anicut (weir) system. The rest of the area falls under

Figure 2 | Proposed irrigation system used for Case 1.

the command areas of 35 tanks and is supplied with water from the tanks, and these are supplemented by the LB canal of the Kaudulla tank to varying degrees depending on the availability of water in the Kaudulla tank.

Accordingly, two cases are considered for the analysis of irrigation water requirements in the LB canal irrigation area:

1. The irrigation demands of the area under minor tanks are met only by the tanks that are under improvement of the system independent of water from the Kaudulla tank, except for the anicut system. Simulation is done considering that the tanks receive inflow from their own catchments.

Five sub-cases are considered for simulation with augmentation of tank capacities under two options as per Table 2 and Figures 1 and 2.

2. The unmet irrigation demands of the area under minor tanks under Case 1 above are also met from the Kaudulla tank as shown in Figure 3.

Both cases are financially analyzed using the net present value method to ascertain the feasibility of the proposals.

The proposal for the new canal system is shown in Table 3 for supplying water to the village tanks to supplement the water release requirements in the season from these tanks. Feeder canals FC1–FC13 are considered for Case 1, and FC1–FC25 are considered for Case 2. Most of the feeder canals are tributaries of the Ambagasoya stream.

RESULTS AND DISCUSSION

The study carried out for the period of 2013–2017 reveals that the water available in the Kaudulla tank, which is fed by its own catchment and from the controlled releases from the Minneriya tank at present, is not sufficient to cultivate 3,790 ha of

Table 1 | Details of minor tanks

No.	Tank name	Coordinates		Capacity (MCM)	Irrigable land (ha)	Catchment area (ha)
		Easting (m)	Northing (m)			
1	Damsopura wewa	231,116	328,505	0.116	371	53
2	Aluthwewa	233,321	323,600	1.111	216	254
3	Illukpitiya wewa	233,787	326,535	0.284	102	93
4	Gaminipura wewa	234,809	326,912	0.289	72	175
5	Rambawewa	233,093	327,614	0.305	88	94
6	Meegaswewa	234,236	332,918	0.261	85	139
7	Pathokwewa	234,214	334,368	0.174	56	53
8	Nikahenawewa	236,371	333,835	0.222	76	202
9	Sathpaththini	234,368	327,228	0.033	22	58
10	Elabatuwewa	237,997	332,665	0.347	78	242
11	Baybiya wewa	231,974	331,352	1.406	128	726
12	Kumbukunawa	235,681	331,505	0.958	71	366
13	Wadiga wewa	235,208	336,386	0.579	187	155
14	Kuda wewa	235,810	332,170	0.044	28	24
15	Badde wewa	234,172	333,500	0.043	15	17
16	Jayagampurawewa	230,154	329,648	0.048	32	44
17	Ekamuthu wewa	230,206	329,119	0.030	25	33
18	Rota wewa	234,569	330,037	0.173	262	177
19	Naripotawewa	236,809	332,975	0.059	32	47
20	Kadurugaswewa	232,616	336,034	0.102	74	95
21	Aruna wewa	234,501	324,480	0.100	159	73
22	Wedikachchiya	225,089	332,873	0.579	48	63
23	Halmillawewa	226,406	333,418	0.740	172	97
24	Weeragollawa wewa	229,735	335,670	0.050	50	70
25	Migollawawewa	231,579	335,525	0.056	97	107
26	Dambagodawela wewa	236,589	326,387	0.180	44	13
27	Palliyagodella wewa	235,476	325,394	0.320	337	73
28	Unakatuwa wewa	237,414	335,471	0.055	12	43
29	Kumarakanda wewa	238,034	336,341	0.060	203	139
30	Gallendiwala wewa	233,346	336,108	0.040	24	110
31	Tank T	221,570	331,824	0.035	26	28
32	Tank U	222,321	333,040	0.045	47	68
33	Tank W	224,551	335,025	0.030	44	107
34	Tank V	226,649	334,823	0.065	49	104
35	Tank S	228,155	335,218	0.055	39	54

paddy area fed by the left bank canal in the Yala season (Figure 4). Enhanced diversion from the Minneriya tank is unlikely in the near future due to the limited water allocation to the tank from the Mahaweli river development scheme.

Therefore, under this situation, it is not realistic to consider feeding the minor tanks from the Kaudulla tank during the water scarce months in the Yala season. Thus, the demand areas of the existing village tanks have to depend on their own catchment yields to cultivate 3,370 ha of land. Different investment options were considered, including the augmentation of eight selected tanks and construction of new canals.

Table 2 | Options for Case 1

Item	Tank name	Capacity (MCM) Existing	Proposed	Potential increase without acquisition of additional lands	Increase through acquisition of additional lands
Option 1	Nikahena	0.222	3	0.44	2.56
	Elabatuwewa	0.347	2.5	0.71	1.79
Option 2	Babiya	1.406	4	1.45	2.55
	Halmiila	0.74	2	0.84	1.16
	Meegaswewa	0.261	1.5	0.42	1.08
	Migollawa	0.056	4	0.24	3.76
	Pathowewa	0.174	1	0.21	0.79
	Wadigewewa	0.579	2	0.74	1.26

Figure 3 | Proposed irrigation system used for Case 2.

Table 3 | Proposed feeder canal system

Feeder canal	Length (m)	From	To
FC1	708	Kudawewa	Naripotawewa
FC2	264	Naripota	Nikahenawewa
FC3	209	Meegaswewa	Baddewewa
FC4	584	Baddewewa	Pathokwewa
FC5	2,726	Rotawewa	Meegaswewa
FC6	148	Kumbukunawawewa	Kudawewa
FC7	1,639	Gaminipurawewa	Dambagodawelawewa
FC8	1,281	Rambawewa	Illukpitiyawewa
FC9	1,143	Rambawewa	Sathpaththiniwewa
FC10	1,015	Illukpitiya wewa	Anurawewa
FC11	1,726	Ambagasoya bifurcation	Migollawa wewa
FC12	365	Rotawewa	Kumbukunawa
FC13	248	Ekamuthuwewa	Jayagampurawewa
FC14	15,108	LB Canal bifurcation	Migollawa wewa
FC15	383	FC 14 bifurcation	Tank T
FC16	746	FC 14 bifurcation	Tank U
FC17	3,544	FC 14 bifurcation	Wedikachchiya wewa
FC18	291	FC 14 bifurcation	Weeragollawa wewa
FC19	434	FC 14 bifurcation	Tank S
FC20	1,232	FC 14 bifurcation	Tank V
FC21	936	FC 14 bifurcation	Tank W
FC22	3,273	LB Canal bifurcation	Damsopura wewa
FC23	2,857	FC 22 bifurcation	Rambawewa
FC24	3,100	FC 23 bifurcation	Rotawewa
FC25	700	FC 22 bifurcation	Ekamuthuwewa

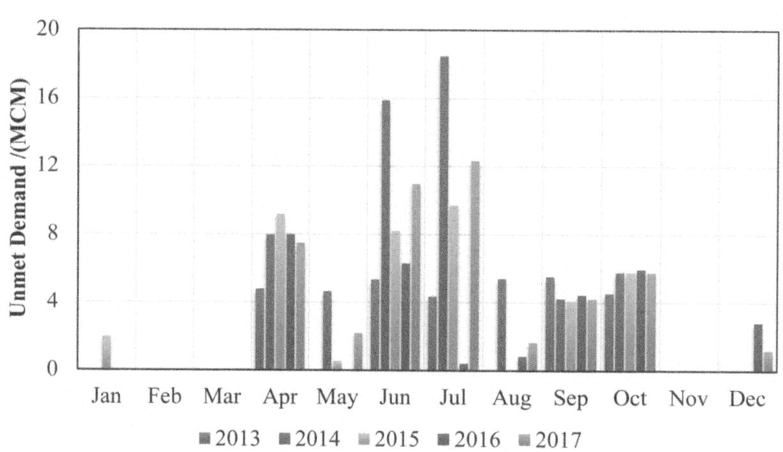

Figure 4 | Unmet water demand under the Kaudulla left bank canal.

WEAP model results

The unmet irrigation demands of the area under minor tanks of the existing system that are not dependent on water from the Kaudulla tank and are instead fed from inflows from their own catchments, for the period 2013–2017, are shown in Figure 5.

The analysis reveals that the annual unmet water demand from 2013 to 2017 varies from 33.75 to 51.86 MCM. The highest unmet demand is shown in the year 2015. Figure 6 shows the monthly unmet demand in 2015.

The maximum monthly unmet water demand of 16.58 MCM is seen in July 2015 of the Yala season.

A proposal for a new canal system is shown in Table 3. Feeder canals FC1–FC13 are considered for Case 1, and FC1–FC25 are considered for Case 2. Most of the feeder canals are tributaries of the Ambagasoya stream.

Figure 7 illustrates the results of the simulation carried out for monthly unmet demand for 2015 with a proposed improvement to the canal system of Case 1-b. The improvement reduces the unmet water demand from 16.58 to 16.35 MCM in July 2015, which is the most water-stressed month of the Yala season.

Figure 8 is the result of a decrease of unmet water demand to 16.35 MCM for the proposed new canal system with the augmentation of tank capacities without inundation of upstream infrastructures as described in Table 4 for July 2015 in the Yala season.

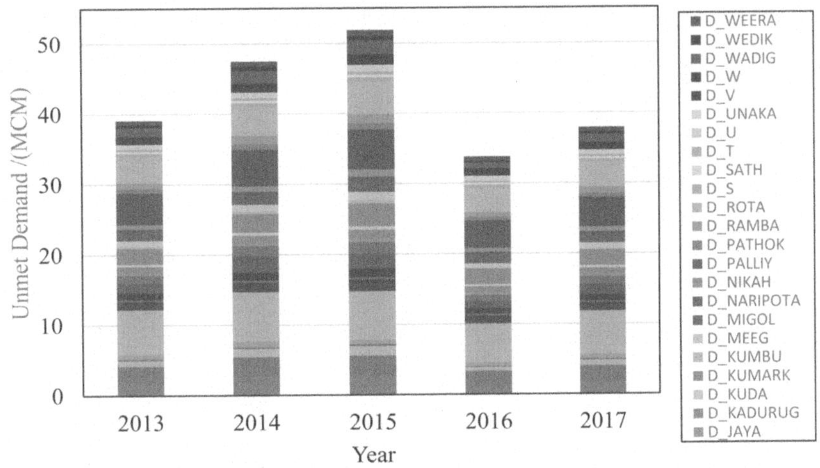

Figure 5 | Unmet demand for Case 1-a from 2013 to 2017.

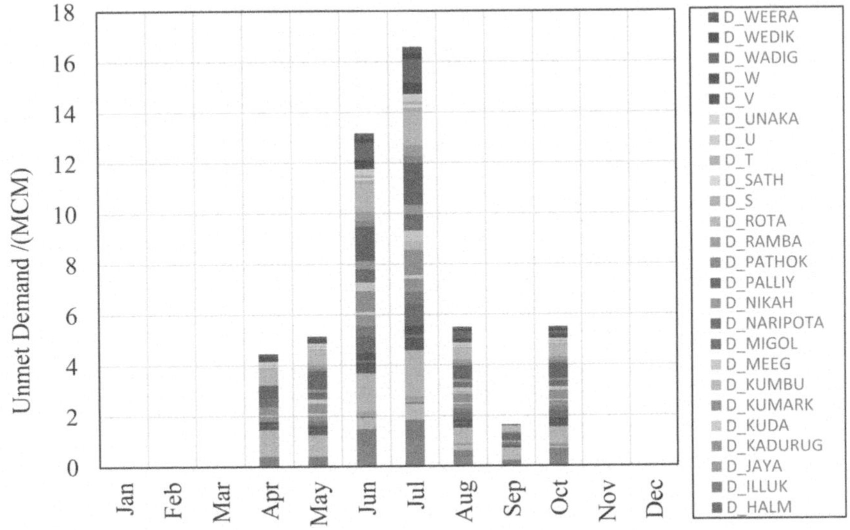

Figure 6 | Monthly unmet demand for Case 1-a for 2015.

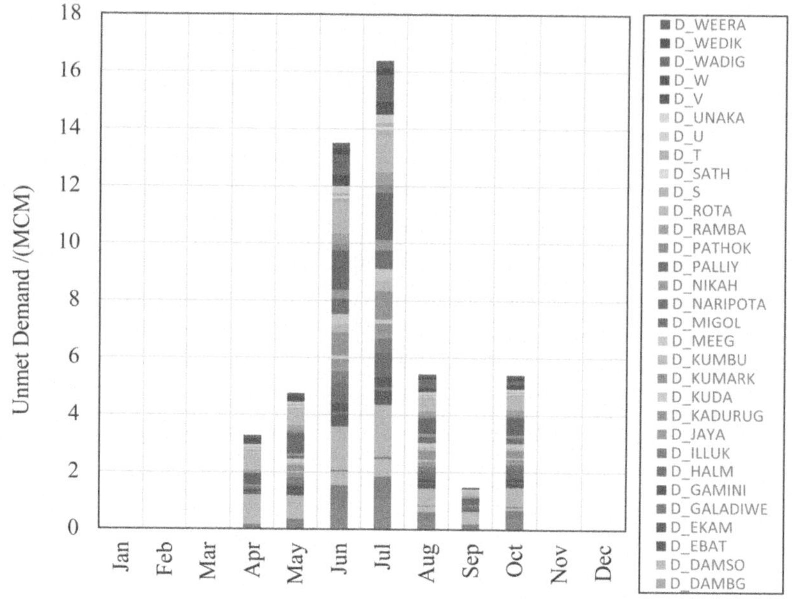

Figure 7 | Monthly unmet demand for Case 1-b for 2015.

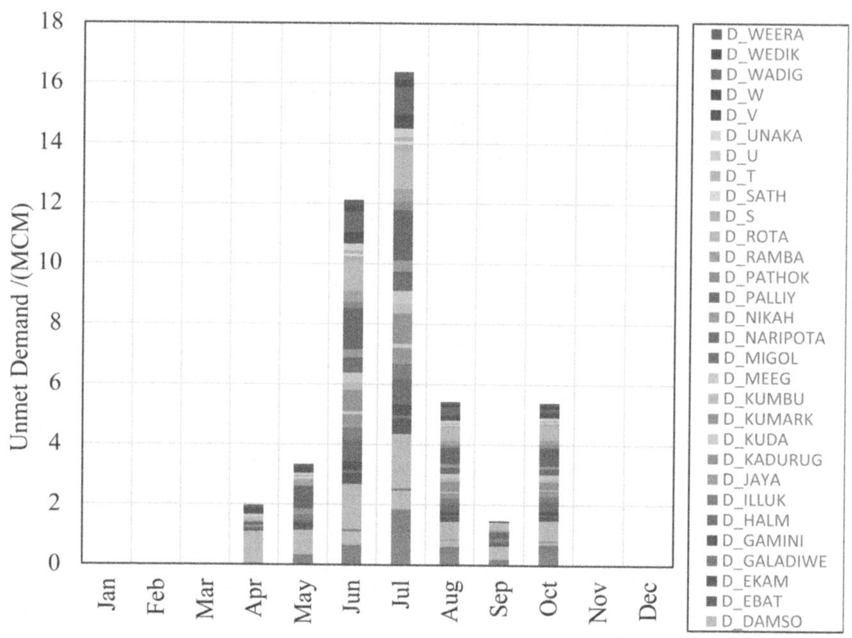

Figure 8 | Monthly unmet demand for Case 1-c for 2015.

Table 4 | Scenarios for Case 1

Case	Description
1-a	Present condition of tanks and canal system as in Figure 1
1-b	Proposed improvement to the canal system as in Figure 2
1-c	1-b and alternate by raising of spill level without inundation of upstream of eight tanks in Table 2
1-d	1-b and augmentation of proposed tank capacities of Option 1 as in Table 2
1-e	1-d and augmentation of proposed tank capacities of Option 2 as in Table 2

Figure 9 shows the decrease in unmet water demand to 14.78 MCM in July 2015 in the Yala season as a result of the proposed new canal system with augmentation of tank capacities of Option 1 in Table 4.

Figure 10 shows a further decrease in unmet water demand to 11.37 MCM in July 2015 in the Yala season because of the proposed new canal system with augmentation of tank capacities of Option 2 in Table 4.

The demand area coverage shown in Figure 11 of 12 selected demand sites varies from 0 to 100% in five cases analyzed in Case 1 for July 2015.

Figure 12 shows the unmet demand for Case 2 for 100% crop intensity.

Case 2 is shown to be the most favorable to reduce the water scarcity in the study area. However, this option is possible only if sufficient water is supplied from the Kaudulla tank to the minor tanks. It is evident from Figure 3(1) that even the supply of irrigation water from the Kaudulla left bank canal to the left bank irrigated area is not fulfilled in the Yala season. Hence, it is

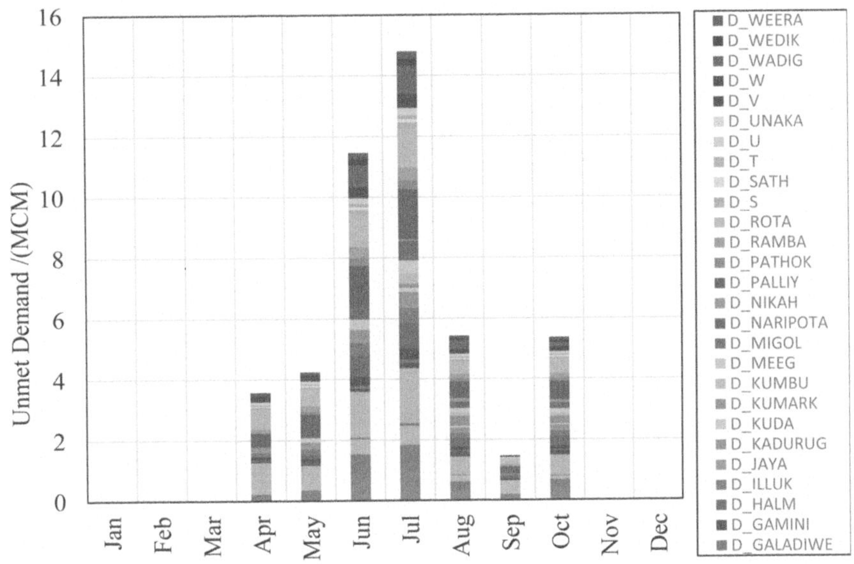

Figure 9 | Monthly unmet demand for Case 1-d for 2015.

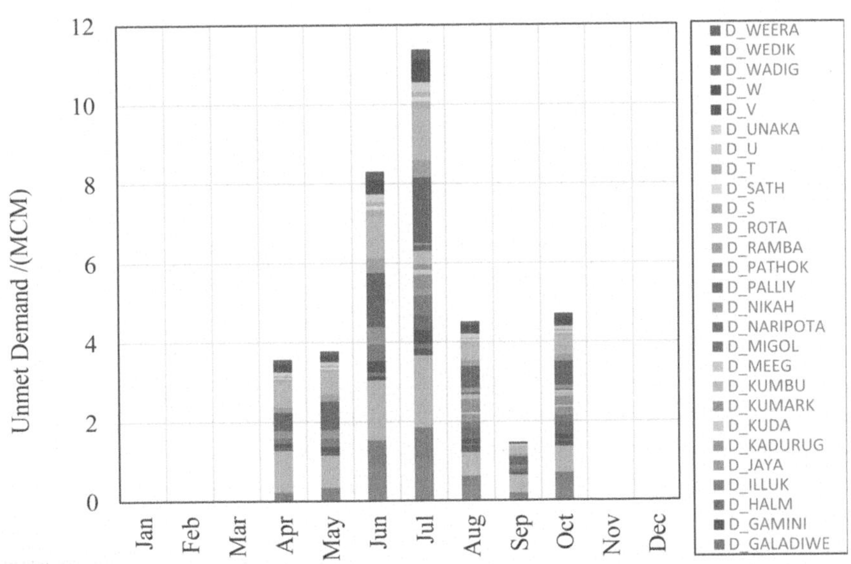

Figure 10 | Monthly unmet demand for Case 1-e for 2015.

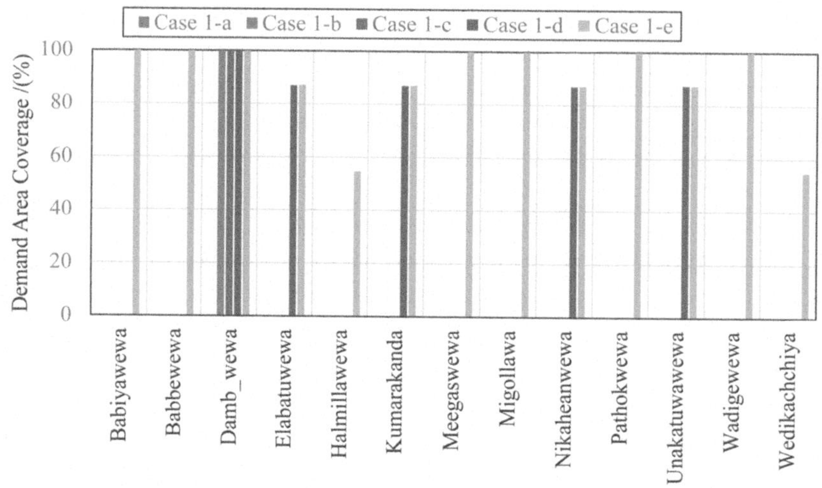

Figure 11 | Demand area coverage for Case 1.

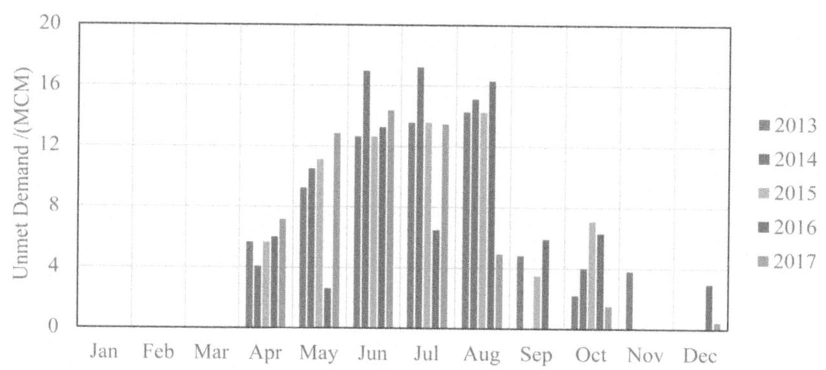

Figure 12 | Unmet demand for Case 2.

not practical to increase the cropping intensity by 100% as described in Case 2, unless there is an enhancement of diversion from Minneriya or Minneriya-Kanthale Yoda Ela to supplement these demands through the Kaudulla tank.

Financial analysis

The following analysis is carried out based on HSR 2016 (Government of Sri Lanka approved schedule of rates –2016) for construction.

Prime cost for the construction of the proposed feeder canal = Rs. 17,120 per meter length
(Capacity: 0.58 m^3/s and Reinforced Cement Concrete-lined (RCC) Canal: 1.2 m × 0.75 m)
Prime cost for augmentation of tank capacity by 1 MCM = Rs. 5,000,000
Prime cost for land acquisition including infrastructures = Rs. 6,000,000/ha
It is assumed that the inflation rate of 7.6% as at 2017 remains unchanged.

Table 5 shows the summary of financial analysis for Case 1.

Payback periods calculated using the net present value by considering 7.6% inflation rate for prime cost and turnover from the yield increment are shown in Table 5. Case 1-e shows that yield can be increased by Rs. 205 million by increasing the total cultivation area in the season through irrigation using minor tanks alone without depending on the water from the left bank canal of Kaudulla tank. That is, an increased area coverage and minimum payback period by augmenting tanks and constructing the proposed canals in Case 1-e.

An economic analysis was carried out using the net present values, and it was found that the option that includes the construction of a new canal system and the augmentation of capacities of village tanks, namely Nikahenawewa, Elabatuwewa,

Table 5 | Summary of cost–benefit for the proposed system

		Prime cost (Rs. million)						
No.	**Description**	**Canal**	**Tanks**	**Land acquisition**	**Total**	**Yield (Rs. million)**	**Yield increment (Rs. million)**	**Payback period using enhancement (years)**
1	Case 1-a					0		
2	Case 1-b	206			206	9	9	>100
3	Case 1-c	206	6		212	9	9	>100
4	Case 1-d	206	25	390	621	71	71	15
5	Case 1-e	206	81	1,266	1,553	205	205	12

Babiyawewa, Halmillawewa, Meegaswewa, Migollawewa, Pathokwewa and Wadigewewa, from the total capacity of 3.8–20 MCM would be the most appropriate option. This intervention (a) increases the income from yield by 205 million with a payback period of 12 years and (b) also increases the area coverage of cultivation under the left bank canal cultivation area of the tank in the Yala season.

CONCLUSIONS

Irrigation water management in the left bank canal development area under the Kaudulla reservoir in the North Central Province of Sri Lanka was analyzed by the application of the HEC-HMS, CROPWAT and WEAP models. The best intervention among a few feasible structural interventions was identified based on the economic return.

Accordingly, the construction of a few new canals and augmentation of capacities of village tanks, namely Nikahenawewa, Elabatuwewa, Babiyawewa, Halmillawewa, Meegaswewa, Migollawewa, Pathokwewa and Wadigewewa tanks, to increase the total capacity from the present 3.8 to 20 MCM was found to be the most appropriate option. This intervention increases the income from the yield by 205 million with a payback period of 12 years.

DATA AVAILABILITY STATEMENT

All relevant data are included in the paper or its Supplementary Information.

REFERENCES

Holmes, D.W., Batchelor, C. H., Gunston, H. & Dawson, R. J. 1978 *Water Management Study at Kaudulla Irrigation Scheme, Sri Lanka*. The Irrigation Department, Sri Lanka.

Jayarathne, J. M. A. R. 2016 *Country Report: Democratic Socialist Republic of Sri Lanka*. Ministry of Disaster Managment, Sri Lanka.

Rekha, Nianthi K. W. G. 2014 Farmers' Responses to Drought: Dry Zone of Sri Lanka: (Case Study in Medirigiriya). *Int. Journal of Case Studies* **5**(6).

First received 28 March 2021; accepted in revised form 22 August 2021. Available online 2 September 2021

© 2022 The Authors Water Management Addressing Societal and Climate Change Challenges doi: 10.2166/wcc.2021.366

Analyzing the duration frequency and severity of drought using copula function in the Yazd city

Mehrtash EskandariPour [a,*] and Shahrokh Soltaninia[b]

[a] Department of Civil Engineering, Isfahan University of Technology (IUT), Isfahan, Iran
[b] Department of Natural Resource Environment, Islamic Azad University, Science and Research Branch, Tehran, Iran
*Corresponding author. E-mail: mehrtashs3@gmail.com

ME, 0000-0001-5990-8361

ABSTRACT

Duration and severity are the two main variables used in drought analysis. The copula functions are appropriate for multivariate drought analysis, as it lacks the limitations of the classical multivariate distribution function. This study investigated the bivariate frequency analysis of drought duration and severity of Yazd city in Iran synoptic station during 1953–2013. To this end, first, the drought duration and severity variables were derived from the 6-month Standardized Precipitation Index. Then, considering the distribution functions, the gamma distribution function was selected for analyzing the severity and the exponential distribution function was selected for analyzing the duration and then the Clayton copula function was selected out of the three copula functions as the most appropriate one. After conducting bivariate frequency analysis, the joint seasonal and conjunctive return period and conditional return period curves were plotted. The current study well signified that multivariate analyses could present better interpretations of the reality; for example, as it was identified in conditional return period curves of the drought, for every constant duration, the amount of the return period changed as the severity changed. On the contrary, in analyzing the univariate of duration, no effects of severity were observed.

Key words: copula function, drought, Iran, joint return period, severity and duration

HIGHLIGHTS

- Use of multivariate analysis in research due to the high degree of correlation between the severity and duration of drought.
- Accurate study of the correlation between the two parameters of hardness and duration of drought.
- Using copula functions in this study that removes the limitations of classical distribution functions.
- Calculation of 6 months.
- Estimation of the parameters of the coupling function.

1. INTRODUCTION

As a natural phenomenon, drought affects wet and dry areas. There are other consequences on the rain-fed agriculture and food security. Different economic sectors could lead to insufficient water resources in cities (Balana *et al.* 2020). Since drought is a critical factor in water resources management and planning (Katipoğlu *et al.* 2020), understanding the characteristics of drought-related problems is an essential step in creating mitigation measures (Bravo *et al.* 2021). Unlike many natural disasters that occur abruptly, drought has a slow pace and is referred to as a creeping phenomenon (Drysdale *et al.* 2020). Moreover, since drought is remarkably associated with unexpected phenomena, such as precipitation and flow patterns (Helama *et al.* 2018), investigating the drought problem requires the use of probability theory and stochastic processes methods. Accordingly, bivariate probabilistic distribution characteristics of drought could be applied which are not derived by a separate variable (Shiau 2006; Zhu *et al.* 2019; Bravo *et al.* 2021). Drought could cost an average of 6–8 billion dollars annually worldwide. Overall, the drought phenomenon also could trigger more damages (Heim 2002; Stone 2010; Sheffield *et al.* 2012). In addition, drought is a complex phenomenon and usually is identified by several correlations. To quantitatively assess the characteristics of drought, one suitable method is using the probability theory and the stochastic process method. However, drought variables, such as duration and severity, are correlated. Univariate analysis of drought characteristics lacks significant detection capability (Wilhite 2000; Huard *et al.* 2006; Kao & Govindaraju 2010; Chen & Sun 2015; Vicente-Serrano *et al.* 2015; She & Xia 2018; Real-Rangel *et al.* 2020). Based on agricultural areas, drought is a costly natural hazard in the world (Wilhite 2000; Heim 2002; Mishra *et al.* 2021). Since droughts are dynamic and multi-index, simultaneous assessment of several drought characteristics is crucial in assessing drought risk (Shiau & Modarres 2009; Zhang & He 2016). Drought is generally referred to as a long-term lack of precipitation. Throughout the history, the drought of typical and frequent climate phenomena has triggered the formation of civilization. Due to negative impacts on the economic, environmental, and social sectors of the system, water delivery design capacity is often employed for urban areas during a critical drought period (Tsiaris *et al.* 2017; Wong 2019). Due to random nature of drought, the best way is to investigate droughts using the probability theory. A significant volume of the research introduced the concept of implementation for drought analysis (Chung & Salas 2000; Shiau & Shen 2001; Gonzalez & Valdes 2003; Tabari *et al.* 2013; Hao & Singh 2015; Tsakiris 2017; Kwon *et al.* 2019; Liu *et al.* 2019; Shamshirband *et al.* 2020; Wu *et al.* 2020).

Drought characteristics have been considered as essential factors in water resources. Complex hydrological phenomena such as droughts and floods are multivariate; thus, several correlated random variables describe them (Wani *et al.* 2019; Wu *et al.* 2020). Two variables of duration and severity are generally applied to describe drought, since the variables have a correlation with each other (Tsakiris *et al.* 2016; Kwon *et al.* 2019; Dai *et al.* 2020). Most of the studies refer to univariate analysis, since the statistical techniques are well developed in hydrology. On the other hand, problems such as limited multivariate statistical techniques, the complexity of mathematical calculations, and the requirement for more data than univariate methods led a few researchers to apply multivariate analyses (Shiau 2006; Azam *et al.* 2018; Ali *et al.* 2020). Many researchers warned that employing univariate analysis, in the presence of the correlation between variables, would lead to overestimated or underestimated results (Salvadori & De Michele 2004; De Michele *et al.* 2005; Nazemi & Elshorbagy 2011; She & Xia 2018; Ben Aissia *et al.* 2017; Wani *et al.* 2019; Aguayo *et al.* 2021). Most of the studies on multivariate analyses of correlated hydrological variables applied the classical multivariate distribution functions such as bivariate normal distribution (Montaseri *et al.* 2018; da Rocha Júnior *et al.* 2020; Li *et al.*, 2021), bivariate exponential distribution (Singh *et al.* 1999; Kao & Govindaraju 2010; Zuo *et al.* 2017; Won *et al.* 2020; Mohsin & Pilz 2021), bivariate gamma distribution (Yue 2001; Vazifehkhah *et al.* 2019; Vergni *et al.* 2020), and bivariate Gumbel distribution (Kim *et al.* 2003; Shiau 2003; Qian *et al.* 2018; Chen *et al.* 2019; Seyedabadi *et al.* 2020). Having fit the bivariate Gumble distribution function on peak discharge and flood runoff volume variables, the conditional return period curves of the peak discharge were drawn for different runoff volumes which were against the return period curve of the marginal distribution of peak discharge (Yang *et al.* 2018; Datta *et al.* 2021). Yue (2001) found that in the presence of positive correlation between two random variables, the conditional return period of a variable was significantly different than its marginal return period. Thus, applying multivariate analyses take the precedence over univariate analyses. For the first time, Shiau (2006) used copula functions for drought analysis. The study fitted copula functions on drought variables of severity and duration and obtained joint return periods of the variables. Applying copula functions in hydrology was first proposed by De Michele *et al.* (2005) through modeling the correlation of precipitation intensity and duration variables. Favre *et al.* (2004) continued using these functions in hydrology. Using multivariate frequency analysis in hydrology was advantageous in terms of copula functions. The classical multivariate probability distribution functions were used in the most attempts to investigate the multivariate analysis of correlated hydrological variables.

The classical multivariate distribution functions have problems, including the specificity of marginal distribution functions, and the determination of their parameters; marginal distribution functions must be of the same type; marginal distribution functions could only be parametric; and classic multivariate distribution functions only use the parametric correlation coefficient and therefore do not consider nonlinear correlations (Karmakar & Simonovic 2009; Chen & Sun 2015). The application of copula function has been presented in hydrology for the first time by modeling the correlation of precipitation duration and intensity variables. Favre *et al.* (2004) continued solving hydrology problems using copula functions. Moreover, correlation modeling of dependent variables by copula functions was introduced in other fields. For example, Bárdossy (2006) found the utility of copula function modeling in identifying groundwater quality specifications. Song & Singh (2010) employed copula functions to analyze precipitation frequency. Samaniego *et al.* (2010) applied copula functions to predict the flow in basins lacking stations. A significant number of studies have been conducted on drought analysis using copula functions (Shiau *et al.* 2007; Serinaldi *et al.* 2009; Song & Singh 2010; Mirabbasi *et al.* 2012; Wong *et al.* 2013; Assani & Guerfi 2017; Kim *et al.* 2018; Han *et al.* 2020; Asif khan *et al.* 2021). One of the statistical analysis problems for drought is determining the correlation between each of the drought parameters or variables. It is well clear that the duration and severity of drought have a mutual relationship, while this is not true for the rain and flood phenomena. Therefore, bivariate analyses are inevitable and copula function could be defined as a function describing the relationship between every related variable. Limitation and problem also could be the inefficiency of the required data for the precise and better calculation of Standardized Precipitation Index (SPI), since the precise daily, long-term and annual rain data might not be available (Serinaldi *et al.* 2009; Shiau & Modarres 2009; Chen *et al.* 2012; Mirabbasi *et al.* 2012; Wong *et al.* 2013; Kwak *et al.* 2014).

In most of the studies investigating drought and the application of copula functions in Iran, the time interval applied for calculating SPI parameters was monthly or seasonal (3 months); however, the current study considered 6 months as the time interval for calculating the SPI of Yazd city. In addition, in most of the studied articles regarding drought and the application of copula functions in Iran, two parameters including severity and duration of drought were utilized for calculating copula functions; however, this study not only used these two essential parameters, but also made the use of frequency parameter in copula function calculations whose similar and simultaneous application of frequency and severity had been less investigated in Iranian articles.

2. MATERIALS AND METHODS

2.1. Drought duration and severity variables

To analyze the frequency, initially, an appropriate probabilistic distribution function should be fitted on variables. Gamma distribution function is generally used in drought studies for checking the severity variable, exponential distribution function, and the duration of the drought (Zhao *et al.* 2017). The same principle was followed in the current article.

Since the maximum likelihood estimation (MLE) method is one of the most widely used parameter estimation methods in hydrology, this method was used in this study, as well. The root-mean-square error (RMSE) is used to select the best distribution function in terms of the appropriateness of fitness, the relation of which is given in Equation (1). It should be noted that p_i and $\hat{p}_i$ are the observable i values and the i values could be calculated from the cumulative distribution function. n is the number of data (Ben Aissia *et al.* 2017; Aguayo *et al.* 2021).

$$\mathrm{RMSE} = \sqrt{\frac{\sum_{i=1}^{n} (p_i - \hat{p}_i)^2}{n}} \tag{1}$$

2.2. Copula functions

2.2.1. Copula function definition

$F_{X,Y}(x,y)$ joint distribution function of correlated random variables of X and Y, which, respectively, have a cumulative distribution of $F_Y(y)$ and $F_X(x)$, can be expressed as Equation (2) according to the theorem (Won *et al.* 2020). $C(u, v)$ is a copula function here. Each copula function has a parameter called the correlation parameter.

$$F_{X,Y}(x, y) = C(F_X(x), F_Y(y)) = C(u, v) \tag{2}$$

The characteristics of the two-parameter gamma functions and the two-parameter exponential function used in the article have been presented in Table 1 (Zuo *et al.* 2017; Mohsin *et al.* 2021).

Table 1 | Probability distribution functions

Probability distribution function	Probability density function	Descriptions
Two-parameter gamma	$f(x) = \frac{x^{\alpha-1}(\exp - x/\beta)}{\beta^{\alpha}\Gamma(\alpha)}$	α: Shape parameter β: scale parameter $\Gamma(.)$: gamma function
Two-parameter exponential	$f(x) = \frac{1}{\alpha}\exp\{-(x-\varepsilon)/\alpha\}$	α: location parameter ε: scale parameter

There are several families of copula functions including the t family: Elliptical, Plackett, Gaussian, and Archimedean. The Archimedean family is employed in numerous studies in water engineering (e.g. Singh *et al.* 2008; Caccamo *et al.* 2011; Hao & Singh 2015; Zuo *et al.* 2017; Chen *et al.* 2019; Vergni *et al.* 2020). Three Archimedean copula functions of Gumbel, Frank, and Clayton were applied in this article to model the correlation of variables. The relationships and characteristics of the three copula functions used in this article have been presented in Table 2, in which u and v denote two dependent cumulative distribution functions which range between 0 and 1; and θ is a parameter used to measure the degree of association between u and v (Tosunoglu & Kisi 2016; Zhang He 2016; Won *et al.* 2020).

2.2.2. Parameter estimation and the fitness test for copula function

There are four main methods to estimate the copula function parameter, regarding which, MLE is the widely applied one (Nazemi & Elshorbagy 2011; Tabari *et al.* 2013; Hao & Singh 2015; Tsakiris *et al.* 2016; Kwon *et al.* 2019). Like the MLE method, this method is also based on maximizing the logarithm of the likelihood function for different values of the correlation parameter, which is given in Equation (3). In the following relation, C is the copula density function. $F_x(x)$ and $F_y(y)$ are marginal cumulative distribution functions and in this citation meaning is for two parameters. To select the best copula function, the higher the MLE logarithm value of a copula function, the more acceptable it is (Tsakiris *et al.* 2016; Datta *et al.* 2021).

$$l(\theta) = \sum_{i=0}^{n} \log[C_\theta(F(X_i), F(Y_i))] \tag{3}$$

2.2.3. Drought joint return periods

According to the studies of Shiau (2006), seasonal and conjunctive overflow probability relationships can be used to obtain joint return periods of drought severity and duration variables through copula function. Also, E(L) is the mean time interval between successive droughts. Equation (4) is used to calculate the conjunctive joint return periods (Yang *et al.* 2018; Datta *et al.* 2021).

$$T^{\wedge}_{D,S}(d, s) = \frac{E(L)}{P(D \geq d \text{ and } S \geq s)} = \frac{E(L)}{1 - F_D(d) - F_S(s) + C(F_D(d), F_S(s))} \tag{4}$$

The joint return period of drought duration (D) and severity (S) could be defined for two cases: the return period for $D \geq d$ and $S \geq s$, and the return period for $D \geq d$ or $S \geq s$, which is denoted by $T^{\wedge}_{D,S}$ and $T^{*}_{D,S}$ Equation (5) is used to calculate the seasonal joint return periods.

$$T^{*}_{D,S}(d, s) = \frac{E(L)}{P(D \geq d \text{ or } S \geq s)} = \frac{E(L)}{1 - C(F_D(d), F_S(s))} \tag{5}$$

The drought duration and severity return period can be conditionally expressed in the following two ways. The drought duration return period is obtained from Equation (6), provided that the severity of the drought exceeds a certain threshold (Kim *et al.* 2018; Han & Singh 2020).

$$T_{D|S \geq s} = \frac{E(L)}{[1 - F_S(s)][1 - F_D(d) - F_S(s) + C(F_D(d), F_S(s))]} \tag{6}$$

where $T_{D|S \geq s}$ denotes the conditional return period for D given $S \geq s$. Drought severity return period is obtained from Equation (7), provided that the duration of the drought exceeds a certain threshold. $T_{S|D \geq d}$ denotes the conditional return

Table 2 | Copula functions

Parameter space	$C(u, v)$	Copula function
$0 < \theta < \infty$	$(u^{-\theta} + v^{-\theta} - 1)^{-1/\theta}$	Clayton
$0 < \theta < \infty$	$\frac{-\log(1 - (1 - e^{-\theta})e^{-t})}{\theta}$	Frank
$1 \leq \theta < \infty$	$\exp(-[(-\ln u)^{\theta} + (-\ln v)^{\theta}]^{1/\theta})$	Gumbel

period for S given D ≥ d (Assani & Guerfi 2017; Kim *et al.* 2018).

$$T_{S|D\geq d} = \frac{E(L)}{[1 - F_D(d)][1 - F_D(d) - F_S(s) + C(F_D(d), F_S(s))]} \tag{7}$$

2.2.4. Standardized Precipitation Index

In the study, the 6-month SPI was employed to calculate drought variables. This index aims to define and monitor the drought. SPI calculation based on long-term precipitation series is applied for a specific period, such as 1, 3, 6, and 12 months (Heim 2002; Stone 2010; Assani & Guerfi 2017). To calculate a 6-month SPI, first, a new precipitation series must be generated from the monthly precipitation series. According to these series, the amount of precipitation per month is obtained from the sum of values of the previous 6 months' precipitation and the same month's precipitation. Then, the empirical Green's function could be fitted on these series of data. Gamma distribution was applied for the severity parameter, and exponential distribution was used for the duration parameter. Then, the relevant parameters of each were calculated. Finally, the inverse of the standard normal distribution function was applied for the calculation of probability values, resulting in a 6-month SPI (Zhao *et al.* 2017; Asif Khan *et al.* 2021).

It should be noted that drought is referred to as a period in which SPI is continuously negative. In other words, SPI lower than zero is considered as drought. Accordingly, the duration of the period is the drought duration, and the sum of the SPI values of the period is the drought severity (Hao & Singh 2015; Wu *et al.* 2021).

3. STUDY DATA

In this study, precipitation data of the Yazd in Iran (Figure 1) synoptic station from 1953 to 2013 was applied. At the 2011 census, the population was 529,673, and it is currently the 15th largest city in Iran. This station has the geographical coordinates of 54.29° longitude, 31.90° latitude, and 1237 altitude meters above the sea level. The average temperature is 19.8 °C,

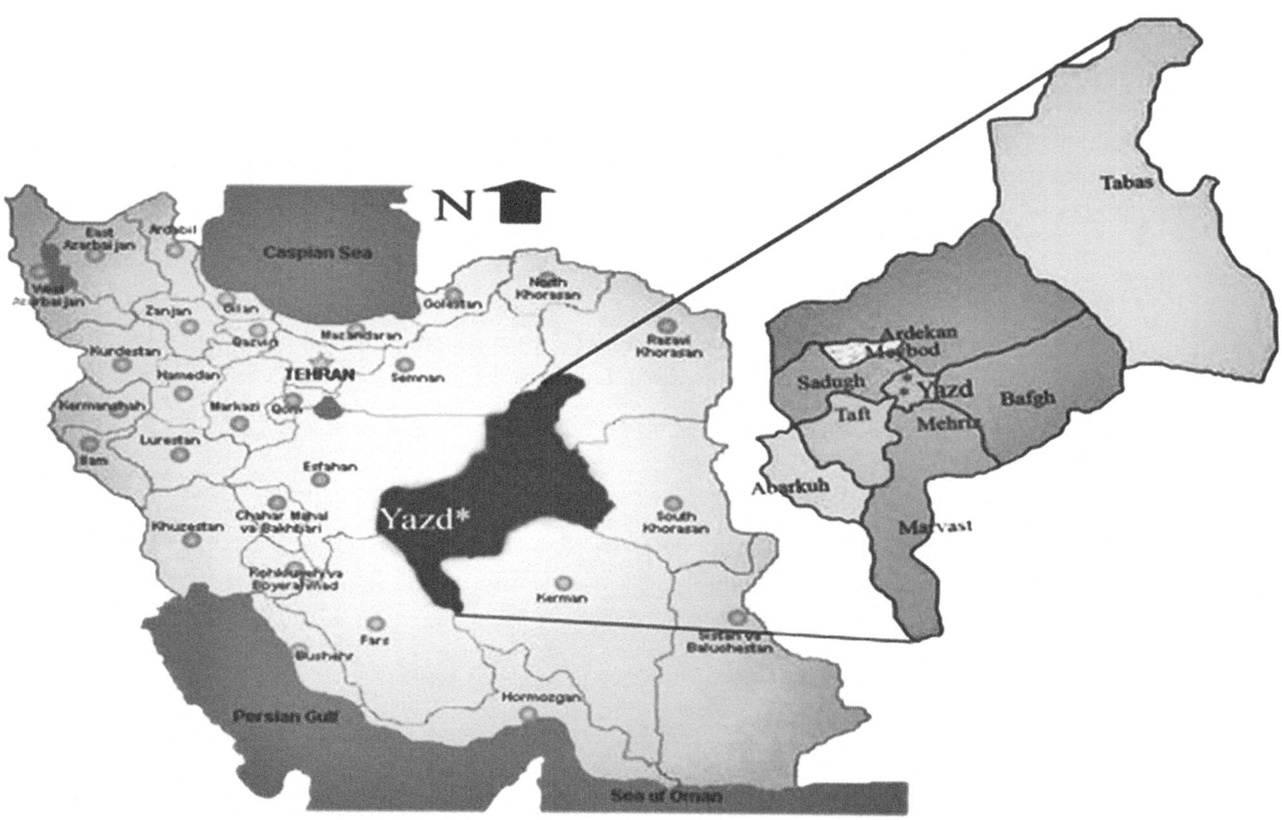

Figure 1 | Map of Iran indicating the location of Yazd County situated in the center of Yazd Province (Dehghan *et al.* 2010).

and the average relative humidity is 31%. The average annual precipitation during the studied years is 46.2 mm. Due to long-term daily statistics, the data validity of the station is high. In recent years, drought triggered costly damages to the lives of the people of Yazd, both in terms of food threats and water shortages. It also had damages to historical and important areas of the Yazd city; thus, considering the problem of drought in Yazd is of utmost importance.

4. RESULTS

4.1. Six-month SPI and drought severity and duration variables

The 6-month SPI series has been shown in Figure 2. In the study, SPI lower than zero is considered as drought, and 45 droughts over a 6-month period during 1953–2013 have been identified by the SPI index.

The curve of severity versus duration for the 6-month SPI is shown as a slip curve in the diagram of Figure 3. This diagram showed that by increasing the duration of drought, the density of drought severity decreased; however, the severity of each parameter increased.

Correlation coefficients between drought severity and the duration of Pearson, Spearman, and Kendall were calculated. Their values are displayed in Table 3 to assess the correlation between drought severity and duration variables. All three correlation coefficients showed significant positive values, so the multivariate analysis was preferable to univariate analysis.

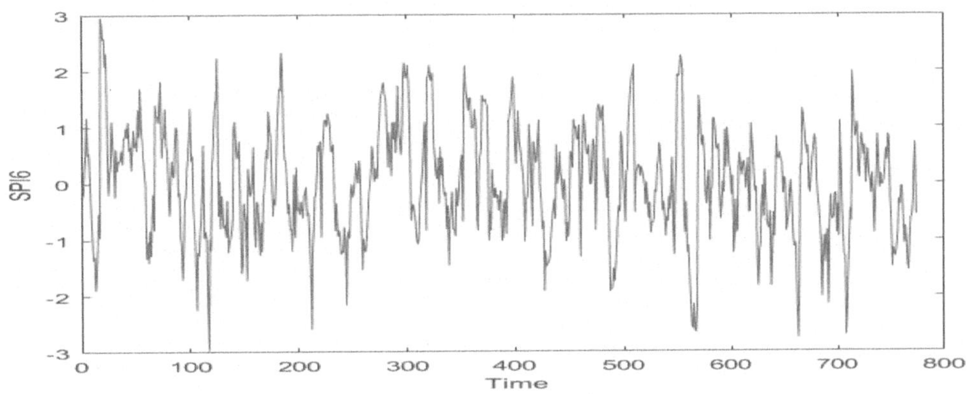

Figure 2 | Six-month SPI for Yazd station.

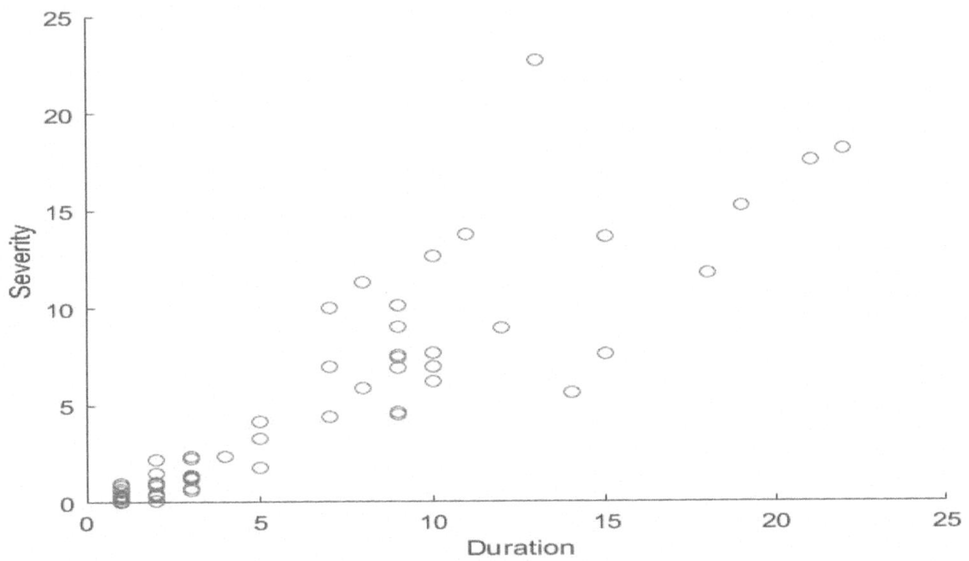

Figure 3 | Severity versus duration of drought for 6-month SPI, which is considered as a slip curve.

Table 3 | Correlation coefficient between drought severity and duration

Correlation coefficient	Spearman	Pearson	Kendall
Value	0.78	0.90	0.71

4.2. Univariate distribution functions

The estimated parameter values of the univariate distribution functions through the two-parameter exponential functions have been presented in the following table regarding which, the duration and the two-parameter gamma functions of severity had the lowest error rates among the relevant functions; the mean squared error value has also been presented accordingly.

The joint cumulative distribution functions of the drought variables estimated by the copula function had been compared to the probability of empirical absence (Table 4). In this study, Green's relations and drawing curves of Figures 4 and 5 were used to obtain the probability of empirical absence. The diagrams showed that as drought severity increased, the probability density decreased; however, the probability of the occurrence of each parameter increased. Green's empirical probability diagram of drought duration and severity is shown in Figures 4 and 5.

4.3. Copula functions

The values of estimated parameters of each of the copula functions using the MLE method and the MLE logarithm are given in Table 5.

According to Table 5, it can be seen that the Clayton copula function had the highest MLE value, which led to its selection as the most suitable copula function. Curves related to any of the calculated copula functions are as follows: Figure 6(a), 6(b) and 6(c) referred

Table 4 | Values of drought severity and duration probability distribution functions

Probability distribution function	Variable	Parameters value	RMSE
Two-Parameter gamma	Drought severity	$\alpha = 0.85$ and $\beta = 18.07$	1.322
Two-parameter Exponential	Drought duration	$\lambda = 5.46$ and $\alpha = 6.23$	1.165

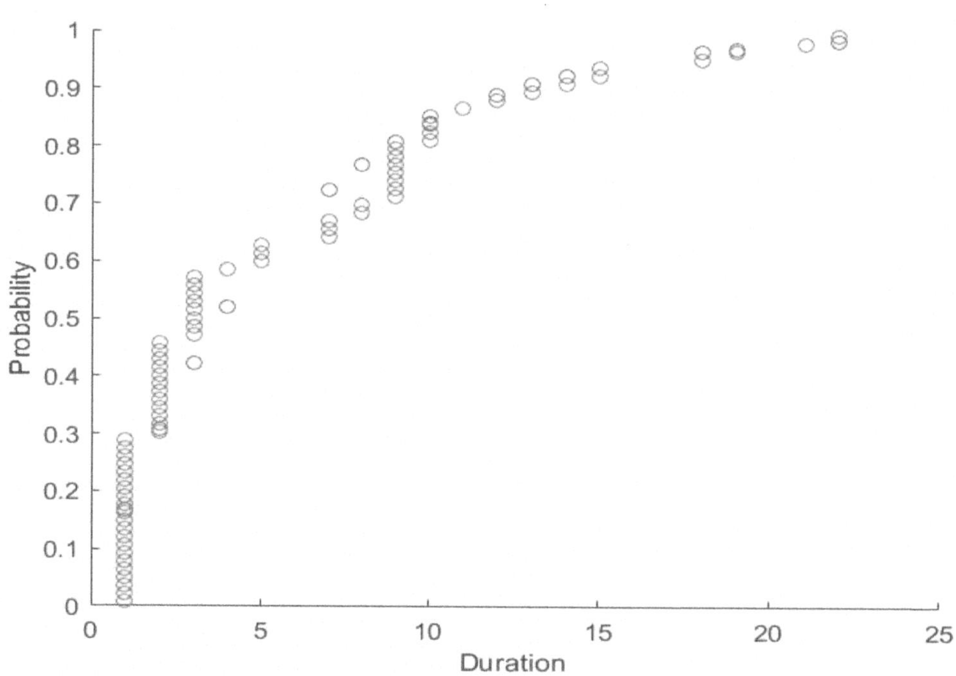

Figure 4 | (Green) empirical probability of drought duration.

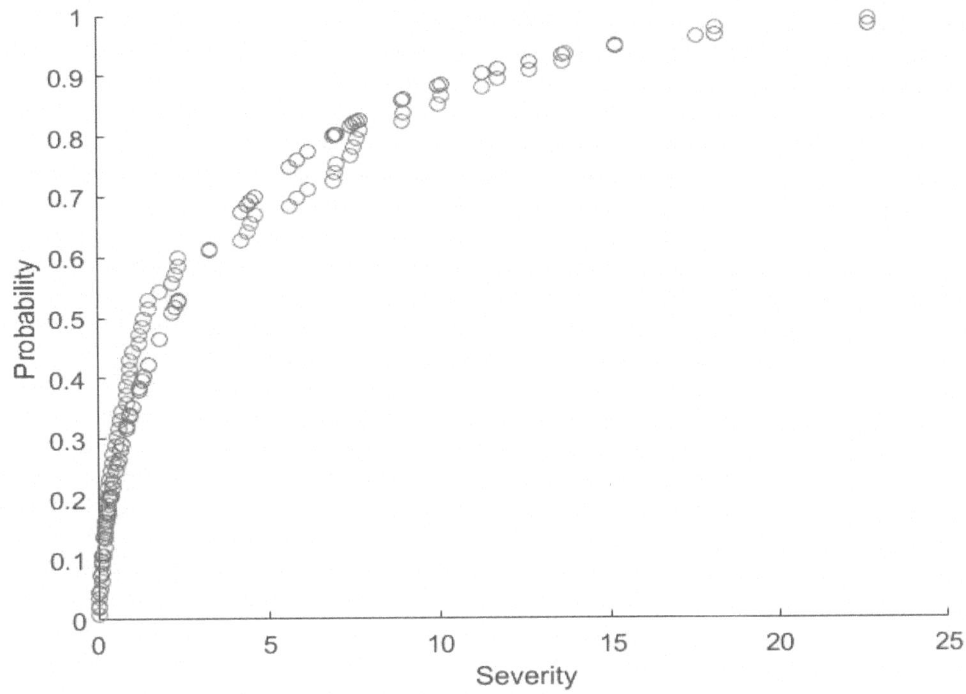

Figure 5 | (Green) empirical probability of drought severity.

Table 5 | Values of copula functions and MLE logarithm

Copula function	Parameter value	MLE
Clayton	3.37	30.02
Frank	16.25	27.16
Gumbel	3.17	29.81

to the Clayton, Frank, and Gumbel, respectively. Given that the drought duration exceeds the 2, 5, 10, 20, 50 and 100-year return period, the variations of the conditional probability density of drought severity for Clayton copula, Frank copula, and Gumbel copula have been shown in Figure 6(a), 6(b) and 6(c). It can be seen that the Clayton copula was suitable for the drought variables.

4.4. Conditional and joint return periods

Equations (4) and (5) are used to calculate seasonal and conjunctive joint return periods (probability return period $P(D \geq d \text{ and } S \geq s)$ and $P(D \geq d \text{ or } S \geq s)$. The results were in the same return period curves as shown in Figure 7(a) and 7(b). The corresponding values were as follows (Table 6).

Figure 7(a) shows that as the monthly time duration increased, the severity value in the type of probability return period decreased. The probability return period was as the following $P(D \geq d \text{ or } S \geq s)$. Figure 7(b) shows that as the monthly time duration increased, the drought severity value in this type of probability return period decreased, and finally, it remained constant. Diagrams of return period drought parameters and duration showed that with increasing the return period, both the amount of drought hardness parameter and the duration of drought duration increased $P(D \geq d \text{ and } S \geq s)$.

Equations (6) and (7) are used to obtain the curves of conditional drought severity and duration return periods. The curves are shown in Figure 8(a) and 8(b). Return period diagrams of severity and duration of drought showed that as the amount of time parameter (D) increased for increasing the amount of severity and as the amount of severity parameter (S) increased for increasing the amount of time parameter, the amount of return periods also increased. The return period of $T(S \mid D \geq d)$ or the duration of 1–10 (Figure 8(a)) and the return period of $T(D \mid S \geq s)$) or the severity of 1–10 (Figure 8(b)) have been shown. Figure 8(a) shows that by increasing the value of severity, the value of the annual return period increased. The conditional

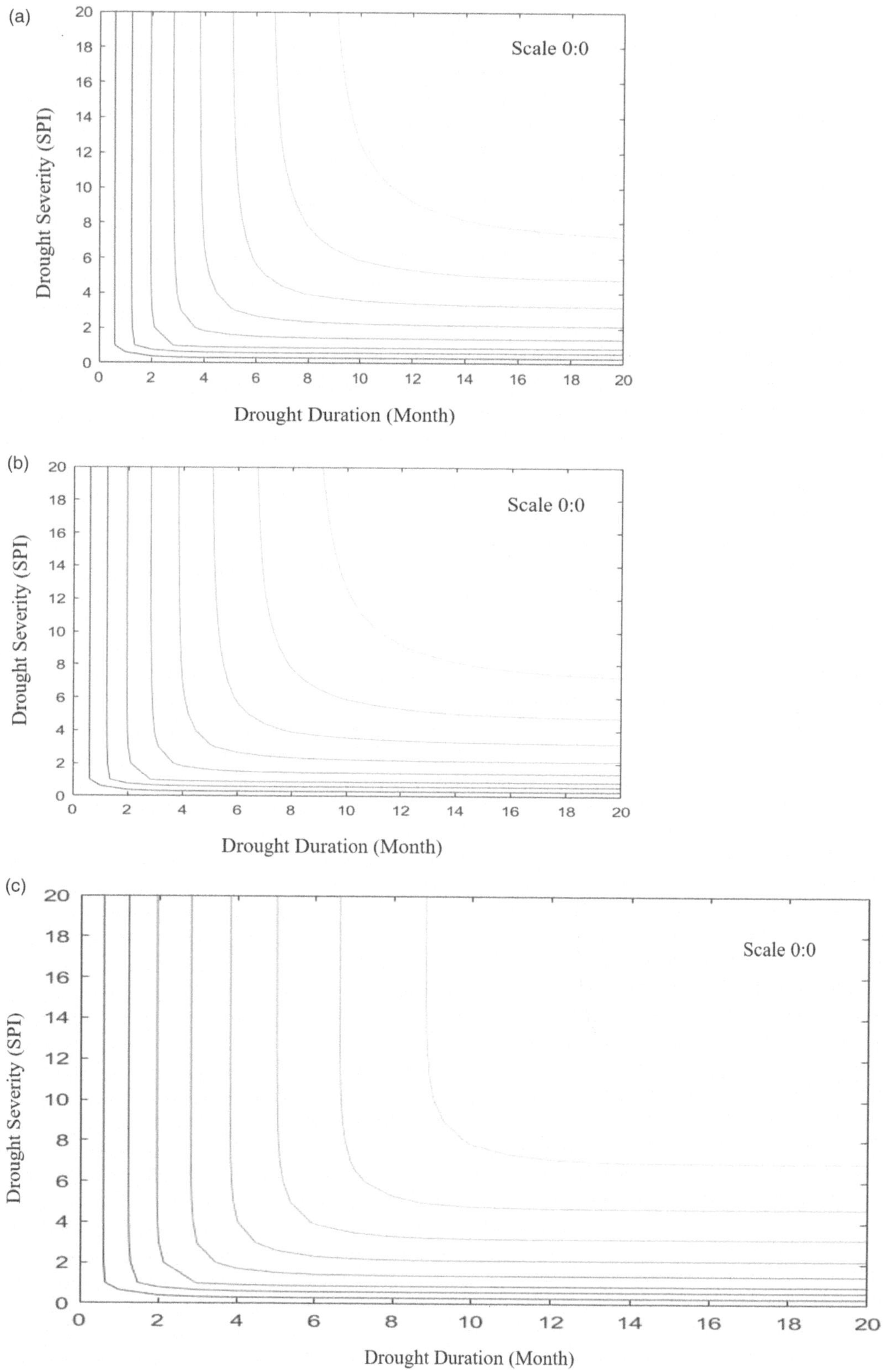

Figure 6 | (a) Clayton severity function based on the monthly time. (b) Frank severity function based on the monthly time. (c) Gumbel severity function based on the monthly time.

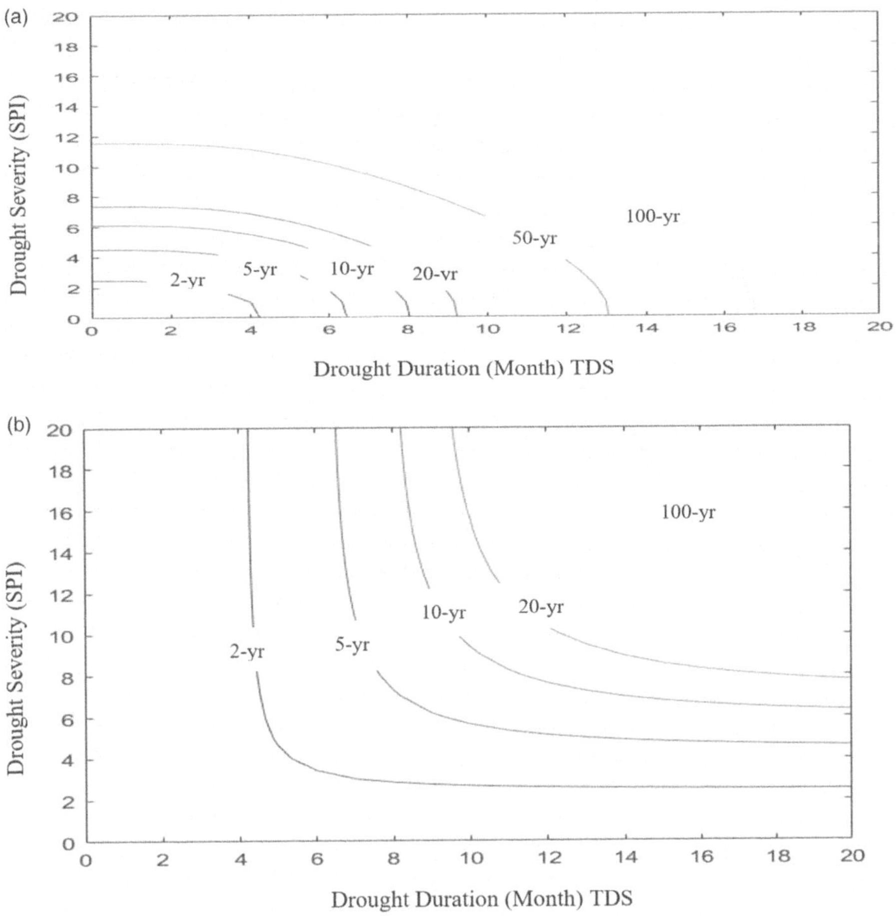

Figure 7 | (a) Contours of joint probabilities of drought severity and duration for Yazd. (b) Contours of joint probabilities of drought severity and duration for Yazd.

Table 6 | Values of severity and duration in proportion to each return period

Severity (SPI)	Duration (months)	Return period (years)
2.48453	4.2421517	2
7.3854142	9.2494869	5
11.566743	13.03739	10
15.960275	16.825293	20
21.971487	21.832628	50
26.623237	25.620531	100

return period of drought severity indicated that the duration was greater than a certain value, D. Figure 8(b) shows that by increasing the duration, the value of the annual return period also increased. The conditional return period of drought duration showed that the severity was greater than a certain value S.

Conditional diagrams of severity and duration of drought showed that the probability of occurrence increases with increasing time parameter for increasing severity and increasing severity parameter for increasing time parameter. Conditional severity and duration diagrams $P(S \leq s \mid D \geq d')$ for the duration of 1–20 and $p(D \leq d \mid S \geq s'$ as well as the severity of 1–20 are as follows. Figure 9(a) shows that by increasing the assumed desired duration, the conditional probability of the curve also increased. This graph is useful for determining the probability of drought severity considering the drought duration exceeding

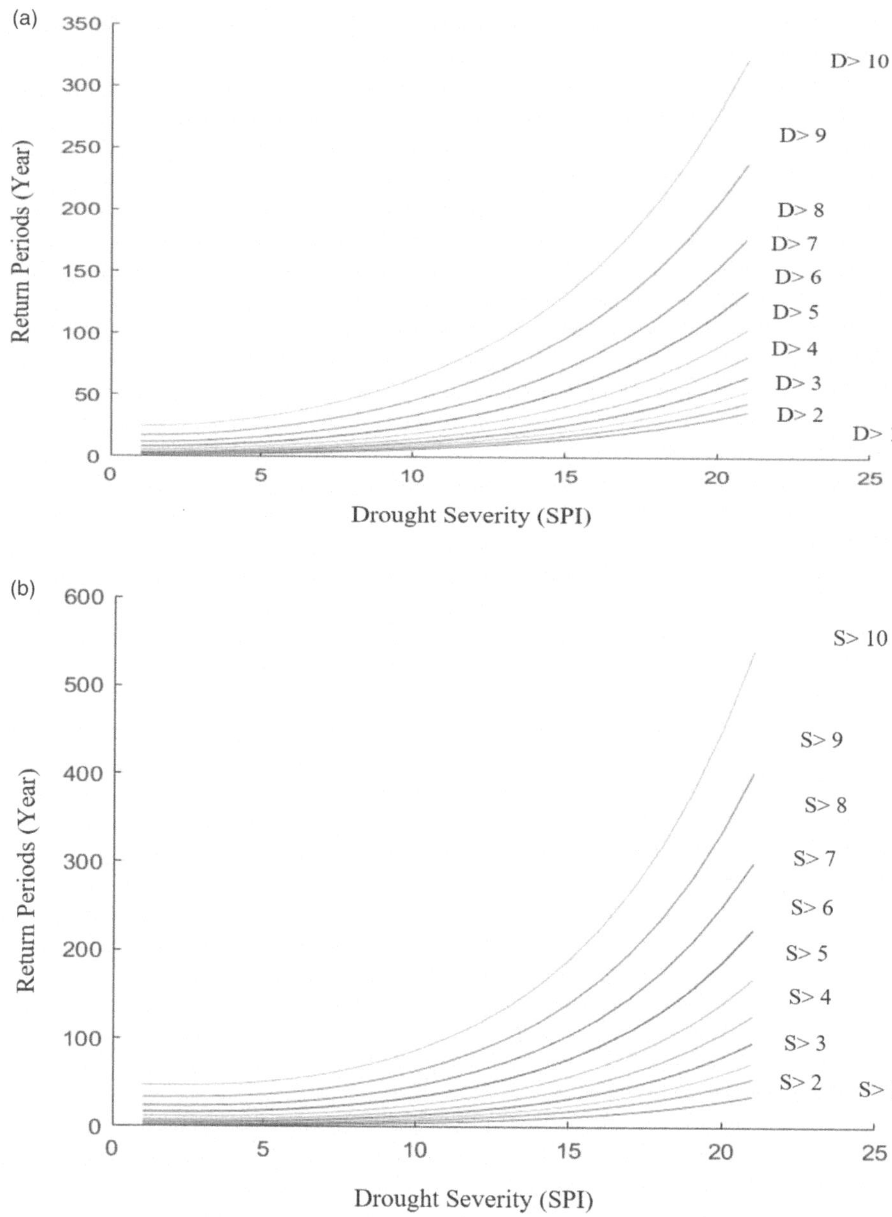

Figure 8 | (a) Joint drought duration and severity return period. (b) Joint drought duration and severity return period.

a certain value. Figure 9(b) shows that by increasing the duration of the desired interval, severity increased the conditional curve.

5. DISCUSSION

The current article has investigated the bivariate frequency analysis of duration and severity of drought as the most significant variables in designing water supply systems. Accordingly, first, each of the drought variables of the 6-month was derived, and the appropriate distribution for each variable was selected among two-parameter gamma functions, type three Pearson, normal and Lognormal three-parameters, two-parameter exponential, Gamble and Wibble three parameters. This study had utilized a two-parameter exponential function for studying the drought duration parameter and had used a two-parameter gamma function for investigating the severity; they were the best parameters in terms of having the least errors. Then, the copula function was selected among Frank, Clayton and Gumbel functions. Clayton function was the best function in this study. Overall, after conducting bivariate frequency analysis, the joint seasonal and conjunctive return period and conditional

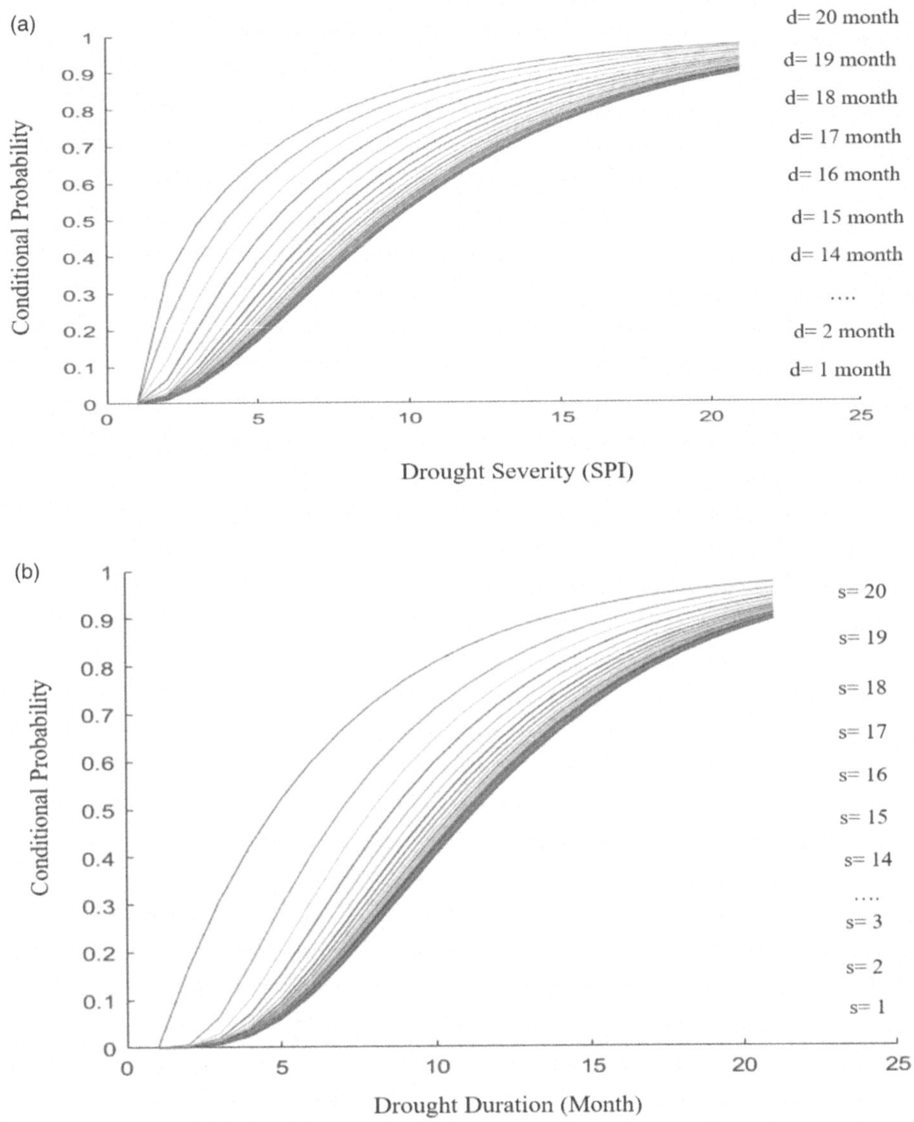

Figure 9 | (a) Conditional return period of drought severity given the drought duration exceeding various values. (b) Conditional return period of drought duration considering drought severity exceeding various values.

return period curves were plotted. The results of multivariate analyses of Copula functions could be used in water resources management plans, risk analysis and probabilistic scheduling. According to the current study, the return period related to any duration and severity could be obtained; for example, the highest observed duration and severity in the studied years equaled 25 months and 2135.19 and the seasonal conjunctive return period of these amounts equaled 1,675 years and 41 years, respectively. Moreover, considering any amounts for the drought severity based on SPI and having the amount of condition for S and D parameters and according to Figure 7(a) and 7(b), the amount of return period could be obtained. For example, according to Figure 7(a), having the severity level equal to 15 and the condition for the duration parameter equal to $D > 6$, the return period amount equaled 75 years.

6. CONCLUSION

According to the study, variables of drought severity and duration had a high correlation. Thus, the multivariate analysis produces presented more reliable results than univariate analysis. In this study, bivariate frequency analysis of drought severity and duration was done using copula functions. Accordingly, first, 6-month SPI series was obtained from the precipitation of Yazd synoptic station during 1953–2013, and the variables of drought severity and duration were derived from this index. The two-parameter gamma functions for severity and the two-parameter exponential functions for the duration were selected as the best functions. The least RMSE was used as the selection criterion. Among the functions of Clayton, Frank, and Gumbel, the Clayton function had the highest value of the MLE logarithm. After conducting bivariate frequency analysis, the joint seasonal and conjunctive return period and conditional return period curves were plotted. The following points could be presented regarding the results of the article:

1. The current study showed that multivariate analysis provided a better analysis of the reality. For example, as can be seen in return period curves during drought duration, as the severity changed, the value of the return period changed for a definite time, whereas in univariate analysis, no severity effects could be observed.
2. Distributions of variables of drought severity and duration were different. Thus, the classical multivariate distribution functions could not be used for multivariate analyses. A copula function is a suitable tool that lacks the problems of the classical multivariate distribution functions.
3. Results of multivariate analyses can be applied in water resources management plans, risk analysis, and probabilistic scheduling. For example, if the design considers a 100-year return period, the various values that make up this return period can be obtained from Figure 7(a) and 7(b). It is also possible to obtain a return period related to any duration and severity.
4. The above results indicated that copula was a useful tool in exploring the associations of the correlated drought variables. However, the current study focused on two random variables as well as one-parameter copulas. Multivariate and two-parameter copula applications could be appropriate topics for future studies.

DATA AVAILABILITY STATEMENT

Data available on request from the authors.

REFERENCES

Aguayo, R., León-Muñoz, J., Garreaud, R. & Montecinos, A. 2021 Hydrological droughts in the southern Andes (40–45°S) from an ensemble experiment using CMIP5 and CMIP6 models. *Sci. Rep.* **11**, 5530. https://doi.org/10.1038/s41598-021-84807.

Ali, Z., Hussain, I., Grzegorczyk, M. A., Ni, G., Faisal, M., Qamar, S., Shokry, A. M., Sharkawy, M. A. V., Gani, S. & Al-Deek, F. F. 2020 Bayesian network based procedure for regional drought monitoring: the seasonally combinative regional drought indicator. *J. Environ. Manag.* **276**, 111296.

Asif Khan, M., Faisal, M., Hashmi, M. Z., Nazeer, A., Ali, Z. & Hussain, I. 2021 Modeling drought duration and severity using two-dimensional copula. *J. Atmos. Sol. Terr. Phys.* **214**, 105530. https://doi.org/10.1016/j.jastp.2020.105530.

Assani, A. A. & Guerfi, N. 2017 Analysis of the joint link between extreme temperatures, precipitation and climate indices in winter in the three hydroclimate regions of Southern Quebec. *Atmosphere* **8**, 75. https://doi.org/10.3390/atmos8040075.

Azam, M., Park, H. K., Maeng, S. J. & Kim, H. S. 2018 Regionalization of drought across South Korea using multivariate methods. *Water* **10**, 24. https://doi.org/10.3390/w10010024.

Balana, B. B., Bizimana, J. C., Richardson, J. W., Lefore, N., Adimassu, Z. & Herbst, B. K. 2020 Economic and food security effects of small-scale irrigation technologies in Northern Ghana. *Water Resour. Econ.* **29**, 100141.

Bárdossy, A. 2006 Copula-based geostatistical models for groundwater quality parameters. *Water Resour. Res.* **42** (11), W11416.

Ben Aissia, M. A., Chebana, F. & Ouarda, T. B. M. J. 2017 Multivariate missing data in hydrology – review and applications. *Adv. Water Resour.* **110**, 299–309. https://doi.org/10.1016/j.advwatres.2017.10.002.

Bravo, R. Z. B., Cunha, A. P. M. d., Leiras, A. & Oliveira, F. L. C. 2021 A new approach for a drought composite index. *Nat. Hazards.* https://doi.org/10.1007/s11069-021-04704-x

Caccamo, G., Chisholm, L. A., Bradstock, R. A. & Puotinen, M. L. 2011 Assessing the sensitivity of MODIS to monitor drought in high biomass ecosystems. *Remote Sens. Environ.* **115**, 2626–2639.

Chen, H. & Sun, J. 2015 Changes in drought characteristics over China using the standardized precipitation evapotranspiration index. *J. Clim.* **28** (13), 5430–5447.

Chen, L., Singh, V. P., Guo, S., Mishra, A. K. & Guo, J. 2012 Drought analysis using copulas. *J. Hydrol. Eng.* **18** (7), 797–780.

Chen, X., Li, F., Li, J. & Feng, P. 2019 Three-dimensional identification of hydrological drought and multivariate drought risk probability assessment in the Luanhe River basin, China. *Theor. Appl. Climatol.* **137**, 3055–3076. https://doi.org/10.1007/s00704-019-02780-5.

Chung, C. H. & Salas, J. D. 2000 Drought occurrence probabilities and risks of dependent hydrologic processes. *J. Hydrol. Eng.* **5** (3), 259–268.

Dai, M., Huang, S., Huang, Q., Leng, G., Guo, Y., Wang, L., Fang, W., Li, P. & Zheng, X. 2020 Assessing agricultural drought risk and its dynamic evolution characteristics. *Agricultural Water Management* **231**, 106003. https://doi.org/10.1016/j.agwat.2020.106003.

da Rocha Júnior, R. L., dos Santos Silva, F. D., Costa, R. L., Gomes, H. B., Pinto, D. D. C. & Herdies, D. L. 2020 Bivariate assessment of drought return periods and frequency in Brazilian northeast using joint distribution by copula method. *Geosciences* **10**, 135. https://doi.org/10.3390/geosciences10040135.

Datta, R., Pathak, A. A. & Dodamani, B. M. 2021 Assessment of meteorological drought return periods over a temporal rainfall change. In: *Trends in Civil Engineering and Challenges for Sustainability. Lecture Notes in Civil Engineering*, Vol. 99 (Narasimhan, M. C., George, V., Udayakumar, G. & Kumar, A., eds). Springer, Singapore. https://doi.org/10.1007/978-981-15-6828-2_42

Dehghan, H., Sadraei, J. & Moosa-Kazemi, H. 2010 The morpological variations of *Culex pipiens* larvae in Yazd Province. *Iran. J. Arthropod Borne Dis.* **4** (2), 42–49.

De Michele, C., Salvadori, G., Canossi, M., Petaccia, A. & Rosso, R. 2005 Bivariate statistical approach to check adequacy of dam spillway. *J. Hydrol. Eng.* **10**(1), 50–57. doi:10.1061/(ASCE)1084-0699(2005)10:1(50).

Drysdale, R. E., Moshabela, M. & Bob, U. 2020 'A creeping phenomenon': the association between rainfall and household food insecurity in the district of iLembe, KwaZulu-Natal. *Clim. Dev.* **10**, 1–11.

Favre, A.-C., El Adlouni, S., Perreault, L., Thiémonge, N. & Bobée, B. 2004 Multivariate hydrological frequency analysis using copulas. *Water Resour. Res.* **40** (1), W01101.

Gonzalez, J. & Valdes, J. B. 2003 Bivariate drought recurrence analysis using tree ring reconstructions. *J. Hydrol. Eng.* **8** (5), 247–258.

Han, E. & Singh, V. P. 2020 Forecasting of droughts and tree mortality under global warming: a review of causative mechanisms and modeling methods. *J. Water Clim. Change* **11** (3), 600–632. https://doi.org/10.2166/wcc.2020.239.

Han, Z., Huang, S., Huang, Q., Bai, Q., Leng, G., Wang, H., Zhao, J., Wei, X. & Zheng, X. 2020 Effects of vegetation restoration on groundwater drought in the Loess Plateau China. *J. Hydrol.* **591** (1), 125566.

Hao, Z. & Singh, V. P. 2015 Drought characterization from a multivariate perspective: a review. *J. Hydrol.* **527**, 668–678.

Heim, R. R. 2002 A review of twentieth-century drought indices used in the United States. *Bull. Am. Meteorol. Soc.* **83** (8), 1149–1166.

Helama, S., Sohar, K., Läänelaid, A. & Jaaguz, J. 2018 Reconstruction of precipitation variability in Estonia since the eighteenth century, inferred from oak and spruce tree rings. *Clim. Dyn.* **50**, 4083–4101. https://doi.org/10.1007/s00382-017-3862-z.

Huard, D., Evin, G. & Favre, A.-C. 2006 Bayesian copula selection. *Comput Stat Data Anal* **51** (2), 809–822

Kao, S.-C. & Govindaraju, R. S. 2010 A copula-based joint deficit index for droughts. *J. Hydrol.* **380** (1), 121–134. https://doi.org/10.1016/j.jhydrol.2009.10.029.

Karmakar, S. & Simonovic, S. P. 2009 Bivariate flood frequency analysis, part 2: a copula-based approach with mixed marginal distributions. *J. Flood Risk Manag.* **2** (1), 32–44.

Katipoğlu, O. M., Acar, R. & Şengül, S. 2020 Comparison of meteorological indices for drought monitoring and evaluating: a case study from Euphrates basin, Turkey. *J. Water Clim. Change* **11** (S1), 29–43. https://doi.org/10.2166/wcc.2020.171.

Kim, T. W., Valdés, J. B. & Yoo, C. 2003 Nonparametric approach for estimating return periods of droughts in arid regions. *J. Hydrol. Eng.* **8** (5), 237–246.

Kim, K., Lee, S. & Jin, Y. 2018 Forecasting quarterly inflow to reservoirs combining a copula-based Bayesian network method with drought forecasting. *Water* **10**, 233. https://doi.org/10.3390/w10020233.

Kwak, J., Kim, D., Kim, S., Singh, V. P. & Kim, H. 2014 Hydrological drought analysis in Namhan river basin, Korea. *J. Hydrol. Eng.* **19** (8), 05014001.

Kwon, M., Kwon, H.-H. & Han, D. 2019 Spatio-temporal drought patterns of multiple drought indices based on precipitation and soil moisture: a case study in South Korea. *Int. J. Climatol.* **39**, 4669–4687. https://doi.org/10.1002/joc.6094.

Li, Y., Gong, Y. & Huang, C. 2021 Construction of combined drought index based on bivariate joint distribution. *Alexandria Engineering Journal* **60** (3), 2825–2833.

Liu, R., Li, Y., Wang, Y., Ma, J. & Cieraad, E. 2019 Variation of water use efficiency across seasons and years: different role of herbaceous plants in desert ecosystem. *Sci. Total Environ.* **647**, 827–835. https://doi.org/10.1016/j.scitotenv.2018.08.035.

Mirabbasi, R., Fakheri-Fard, A. & Dinpashoh, Y. 2012 Bivariate drought frequency analysis using the copula method. *Theor. Appl. Climatol.* **108** (1–2), 191–206.

Mishra, A., Bruno, E. & Zilberman, D. 2021 Compound natural and human disasters: managing drought and COVID-19 to sustain global agriculture and food sectors. *Sci. Total Environ.* **754**, 142210. https://doi.org/10.1016/j.scitotenv.(2020).142210.

Mohsin, M. & Pilz, J. 2021 Stochastic model for drought analysis of the Colorado River Basin. *Stoch. Environ. Res. Risk Assess.* https://doi.org/10.1007/s00477-021-01989-z.

Montaseri, M., Amirataee, B. & Rezaie, H. 2018 New approach in bivariate drought duration and severity analysis. *J. Hydrol.* **559**, 166–181.

Nazemi, A. & Elshorbagy, A. 2011 Application of copula modeling to the performance assessment of reconstructed watersheds. *Stoch. Environ. Res. Risk Assess.* **26** (2), 189–205.

Qian, L., Wang, H., Dang, S., Wang, C., Jiao, Z. & Zhao, Y. 2018 Modelling bivariate extreme precipitation distribution for data-scarce regions using Gumbel–Hougaard copula with maximum entropy estimation. *Hydrol. Process.* **32**, 212–227. https://doi.org/10.1002/hyp.11406.

Real-Rangel, R. A., Pedrozo-Acuña, A., Breña-Naranjo, J. A. & Alcocer-Yamanaka, V. H. 2020 A drought monitoring framework for data-scarce regions. *J. Hydroinform.* **22** (1), 170–185. https://doi.org/10.2166/hydro.2019.020.

Salvadori, G. & De Michele, C. 2004 Frequency analysis via copulas: theoretical aspects and applications to hydrological events. *Water Resour. Res.* **40** (12), W12511.

Samaniego, L., Bárdossy, A. & Kumar, R. 2010 Streamflow prediction in ungauged catchments using copula-based dissimilarity measures. *Water Resour. Res.* **46** (2), W02506.

Serinaldi, F., Bonaccorso, B., Cancelliere, A. & Grimaldi, S. 2009 Probabilistic characterization of drought properties through copulas. *Phys. Chem. Earth* **34** (10–12), 596–605.

Seyedabadi, M., Kavianpour, M. & Moazami, S. 2020 Multivariate drought risk analysis based on copula functions: a case study. *Water Supply* **20** (6), 2375–2388. https://doi.org/10.2166/ws.2020.153.

Shamshirband, S., Hashemi, S., Salimi, H., Samadinfard, S., Asadi, E., Shadkani, S., Kargar, K., Mosavi, A., Nabipour, N. & Chau, K.-W. 2020 Predicting standardized streamflow index for hydrological drought using machine learning models. *Engineering Applications of Computational Fluid Mechanics* **14** (1), 339–350. doi.10.1080/19942060.2020.1715844.

She, D. & Xia, J. 2018 Copulas-based drought characteristics analysis and risk assessment across the Loess Plateau of China. *Water Resour. Manage.* **32**, 547–564. https://doi.org/10.1007/s11269-017-1826-z.

Sheffield, J., Wood, E. F. & Roderick, M. L. 2012 Little change in global drought over the past 60 years. *Nature* **491** (7424), 435–438.

Shiau, J. T. 2003 Return period of bivariate distributed hydrological events. *Stoch. Environ. Res. Risk Assess.* **17** (1–2), 42–57.

Shiau, J. T. 2006 Fitting drought duration and severity with two-dimensional copulas. *Water Resour. Manag.* **20** (5), 795–815.

Shiau, J. T. & Modarres, R. 2009 Copula-based drought severity-duration-frequency analysis in Iran. *Meteorol. Appl.* **16**, 481–489.

Shiau, J. T. & Shen, H. W. 2001 Recurrence analysis of hydrologic droughts of differing severity. *J. Water Resour. Plan. Manag.* **127** (1), 30–40.

Shiau, J. T., Feng, S. & Nadarajah, S. 2007 Assessment of hydrological droughts for the Yellow River, China, using copulas. *Hydrol. Process.* **21** (16), 2157–2163.

Singh, B. B., Mai-Kodomi, Y. & Terao, T. 1999 A simple screening method for drought tolerance in cowpea. *Indian J. Genet.* **59**, 211–220.

Singh, G. P., Chaudhary, H. B., Rajbir, Y. & Tripathi, S. 2008 Genetic analysis of moisture stress tolerance in segregating populations of bread wheat (*T. aestivum* L.). *Indian J. Agric. Sci.* **78** (10), 848–852.

Song, S. & Singh, V. P. 2010 Frequency analysis of droughts using the Plackett copula and parameter estimation by genetic algorithm. *Stoch. Environ. Res. Risk Assess.* **24**, 783–805.

Stone, R. 2010 Severe drought puts a spotlight on Chinese dams. *Science* **327** (5971), 1311–1311.

Tabari, H., Nikbakht, J. & Hosseinzadeh Talaee, P. 2013 Hydrological drought assessment in Northwestern Iran based on Streamflow Drought Index (SDI). *Water Resour. Manag.* **27**, 137–151. https://doi.org/10.1007/s11269-012-0173-3.

Tosunoglu, F. & Kisi, O. 2016 Joint modeling of annual maximum drought severity and corresponding duration. *J. Hydrol.* **543** (Part B), 406–422.

Tsakiris, G. 2017 Drought risk assessment and management. *Water Resour. Manag.* **31**, 3083–3095. https://doi.org/10.1007/s11269-017-1698-2.

Tsakiris, G., Kordalis, N., Tigkas, D., Tsakiris, V. & Vangelis, H. 2016 Analysing drought severity and areal extent by 2D Archimedean copulas. *Water Resour. Manag.* **30**, 5723–5735.

Tsiaras, K. P., Christodoulaki, S., Petihakis, G., Frangouli, C. & Triantafyllou, G. 2017 Model simulations of a mesocosm experiment investigating the response of a low nutrient low chlorophyll (LNLC) marine ecosystem to atmospheric deposition events. *Front. Mar. Sci.* **4**, 120. doi:10.3389/fmars.2017.00120.

Vazifehkhah, S., Tosunoglu, F. & Kahya, E. 2019 Bivariate risk analysis of droughts using a nonparametric multivariate standardized drought index and copulas. *J. Hydrol. Eng.* **24** (15). https://doi.org/10.1061/(ASCE)HE.1943-5584.0001775

Vergni, L., Todisco, F., Di Lena, B. & Mannocchi, F. 2020 Bivariate analysis of drought duration and severity for irrigation planning. *Agricul. Water Manag.* **229**, 105926. https://doi.org/10.1016/j.agwat.2019.105926.

Vicente-Serrano, S. M., Van der Schrier, G., Beguería, S., Azorin-Molina, C. & Lopez-Moreno, J-I. 2015 Contribution of precipitation and reference evapotranspiration to drought indices under different climates. *J. Hydrol.* **526**, 42–54.

Wani, O., Scheidegger, A., Cecinati, F., Espadas, G. & Rieckermann, J. 2019 Exploring a copula-based alternative to additive error models for non-negative and autocorrelated time series in hydrology. *J. Hydrol.* **575**, 1031–1040.

Wilhite, D. A. 2000 Drought as a natural hazard: concepts and definitions. In: *Drought: A Global Assessment* (Wilhite, D. A., ed.). Routledge, New York.

Won, J., Choi, J., Lee, O., Park, M. J. & Kim, S. 2020 Two ways to quantify Korean drought frequency: partial duration series and bivariate exponential distribution, and application to climate change. *Atmosphere* **11**, 476. https://doi.org/10.3390/atmos11050476.

Wong, A. 2019 An assessment of drought on maize cropping success in ancient Maya lowlands during the last half of the first millennium ce. *J. Water Clim. Change* **10** (3), 546–553. doi:https://doi.org/10.2166/wcc.2018.180.

Wong, G., van Lanen, H. A. J. & Torfs, P. J. J. F. 2013 Probabilistic analysis of hydrological drought characteristics using meteorological drought. *Hydrol. Sci. J.* **58** (2), 253–270.

Wu, B., Ma, Z. & Yan, N. 2020 Agricultural drought mitigating indices derived from the changes in drought characteristics. *Remote Sens. Environ.* **244**, 111813.

Wu, L., Chang, Y., Wang, L., Wang, S. & Wu, J. 2021 Genetic dissection of drought resistance of common bean at the seedling stage by genome-wide association study. *Research Square* https://doi.org/10.21203/rs.3.rs-752443/v1

Yang, J., Chang, J., Wang, Y., Li, Y., Hu, H., Chen, Y., Huang, Q. & Yao, J. 2018 Comprehensive drought characteristics analysis based on a nonlinear multivariate drought index. *J. Hydrol.* **557**, 651–667.

Yue, S. 2001 A bivariate extreme value distribution applied to flood frequency analysis. *Nord. Hydrol.* **32** (1), 49–64.

Zhang, B. & He, C. 2016 A modified water demand estimation method for drought identification over arid and semiarid regions. *Agric. Meteorol.* **230–231**, 58–66.

Zhao, P., Lü, H., Fu, G., Zhu, Y., Su, J. & Wang, J. 2017 Uncertainty of hydrological drought characteristics with copula functions and probability distributions: a case study of Weihe River, China. *Water* **9**, 334. https://doi.org/10.3390/w9050334.

Zhu, Y., Liu, Y., Singh, W. V. P., Ma, X. & Yu, Z. 2019 Three dimensional characterization of meteorological and hydrological droughts and their probabilistic links. *J. Hydrol.* **578**, 124016. https://doi.org/10.1016/j.jhydrol.2019.124016.

Zuo, D., Hou, W. & Hu, J. 2017 An entropy-based investigation into bivariate drought analysis in China. *Water* **9**, 632. https://doi.org/10.3390/w9090632.

First received 28 December 2020; accepted in revised form 11 August 2021. Available online 1 September 2021

© 2022 The Authors Water Management Addressing Societal and Climate Change Challenges doi: 10.2166/wcc.2021.079

Governance conundrum in pursuit of the human right to water and sanitation: tracking the progress of the leave-no-one-behind principle in Uganda

G. Z. Ahabwe[a,*], D. W. Batega[a], A. Ssewaya[a] and C. B. Niwagaba[b]

[a] Department of Sociology, Makerere University, P.O Box 7062 Kampala
[b] Department of Civil and Environmental Engineering, Makerere University, P.O Box 7062 Kampala
*Corresponding author. E-mail: gzahabwe@chuss.mak.ac.ug

GZA, 0000-0002-1521-6424; CBN, 0000-0003-1494-1939

ABSTRACT

The Government of Uganda is a party to numerous international and national commitments, laws and regulatory frameworks to deliver the socio-economic transformation of its citizens. Sustainable Development Goals (SDGs) commit states, Uganda inclusive, to achieve a better and more sustainable future for all by 2030. According to the SDG 6 resolutions, the rights to water and sanitation are legally binding upon states, which have the primary responsibility to ensure their full realization, including unserved and under-served areas. While the SDGs came into force after 2015, it is not clear the extent to which Uganda's policies, legal instruments and practice are aligned to the SDGs. A combined methodology was adopted to undertake the study under a cross-sectional design. The study reviewed relevant literature such as sector performance reports on water and sanitation. Purposive sampling was used to select relevant informants for primary data. Key informant interviews were held with representatives of relevant government agencies, United Nations agencies, civil society organisations, indigenous communities and the private sector. Findings indicate that while the Government of Uganda has made tremendous efforts regarding the right to water and sanitation, enormous challenges and gaps remain. For instance, the pro-poor strategies to access safe water, sanitation and hygiene facilities are largely insignificant.

Key words: human rights, leave-no-one-behind, sanitation, vulnerable groups, water

HIGHLIGHTS

- This paper shows the evidence that the rights-based approach is not yet understood by many Ugandans.
- This paper highlights the mismatch between policy and practice.
- This research showcases the gaps in data capture to inform effective monitoring and evaluation.
- This research also demonstrates the progress that the Government of Uganda has made in ensuring water access to all.

INTRODUCTION

The United Nations (UN) General Assembly Resolution 70/1 (2015) that adopted the Sustainable Development Goal (SDG) framework affirms, 'these goals seek to realize the human rights of all and to achieve gender equality'. Resolution 70/1 mentions explicitly the Universal Declaration of Human Rights (UDHR). Goal 6 specifically relates to the realization of the Human Rights to Water and Sanitation (HRWS) by calling for the realization of universal and equitable access to safe and affordable drinking water and to adequate and equitable sanitation and hygiene for all by 2030 (Goals 6.1 and 6.2). Uganda is a signatory to this resolution.

Understanding both water and sanitation governance is a challenging exercise, as there are multiple governance processes in which different roles and power relations among different interest groups and stakeholders are rehearsed (Pahl-Wostl 1995). The chain of actors and attendant policies are quite varied from national, at the macro level, to districts at the meso level, which often includes health inspectors and private sector entities, to communities and individual households at the micro level. The HRWS are derived from international instruments to which states have signed up, such as the UN General Assembly Resolution 64/292 (2010) or the UN General Assembly Resolution 74/141. According to these resolutions, the rights to water and sanitation are legally binding upon states, which have the primary responsibility to ensure their full realization, including unserved and under-served areas.

Globally, the HRWS remains far-fetched for many people. While the system of human rights in the UN was introduced by the UDHR in 1948, the HRWS was not explicitly recognized until 28 July 2010 when the UN General Assembly and the Human Rights Council recognized HRWS through Resolution 64/292.

Uganda has adopted a number of instruments, legislation/laws, policies, strategic and development plans, and regulatory frameworks at the national, sub-national/local government, regional and international levels concerning the HRWS in Uganda. International bodies have risen to promote HRWS in Uganda through cooperating with the Ministry of Water and Environment (MWE), umbrella organizations under the ministry as well as non-governmental organizations (NGOs) to help in the supply of water and improving access to sanitation. This has been through funding programmes and projects of water and sanitation. Uganda is a signatory to the UN SDGs regarding water and sanitation, making these programmes part of the international aid in the fight for HRWS, leave-no-one-behind (LNOB) and gender equality. The HRWS have been incorporated in the legislation, national frameworks, strategies, SDGs, policies, projects and programmes at both national and sub-national levels. Uganda is good at policymaking. Indeed, many laws, policies and institutional frameworks are in place, right from the constitution of the Government of Uganda.

In Uganda, categories of left-behind groups or vulnerable populations include the elderly, women, children and students, nomadic populations, extremely poor populations, refugees and internally displaced people, people with physical or developmental disabilities, geographically dispersed or hard-to-reach populations as well as groups excluded based on sexual orientation such as Lesbian, Gay, Bisexual, Transgender and Questioning (or queer) (LGBTQ). This paper examined the link between policy and practice in fulfilling the HRWS and the leave-no-one-behind (LNOB) principle for the vulnerable and marginalized people in Uganda. The purpose of the review was to take stock of the national integration of the HRWS in target countries, identify gaps and challenges, and related recommendations, as well as suggest best practices in 'LNOB'.

Specifically, the study examined the role and responsibilities of various institutions – public and private – including ministries, local governments, water & sanitation supply providers, and water users' associations in the provision of water, sanitation and hygiene (WASH) services. Secondly, the study examined the activities of these institutional actors in the achievement and implementation of HRWS principles (availability, acceptability, accessibility, affordability, quality and safety, dignity and privacy, participatory action, non-discrimination and sustainability). The study assessed whether adequate WASH services are provided in public spaces such as schools and markets and whether these are convenient for both children and the elderly. Thirdly, the study examined the extent of the vulnerable and marginalized people's population's participation (functional, instrumental, consultative and transformational) in decision-making and their contribution to leadership and accountability (in the form of monitoring and evaluation) at national and sub-national levels. Fourthly, it was also important to look at informational asymmetries, thus examining whether information, for example, on women's rights to water, hygiene and sanitation is culturally appropriate and accessible since power asymmetries affect vulnerable groups' ability to seek justice with regard to violations of their HRWS.

The working hypothesis for this study was that harmonizing policy and practice is a prerequisite for Uganda's trajectory to achieving the right to water and sanitation for all. Consequently, this study sought to answer three research questions: what are Uganda's WASH policy and institutional arrangements to operationalize the SDG agenda, what are the current practices in operationalizing the SDG framework and what needs to be done to fast track the realization of the HRWS and the LNOB principle?

METHODS

This study adopted a cross-sectional design to address the research questions. This type of research design selects a representative subset of a given population to collect data about research questions of interest. It is called cross-sectional because the information gathered about the study problem represents a snapshot of what is going on at only one point in time (Olsen & Marie 2004). The purposive sampling technique was adopted to enable the researcher to select respondents who have knowledge and experience about the subject of study. Qualitative data collection was applied and it involved key informant interviews and focus group discussions.

The study adopted the principle of data saturation in determining the key informant sample size. The key WASH service providers in Uganda were listed. Thereafter, data collection started with the key informants from government and development partners. Additional key informants were enrolled until a point was reached where no new information was coming out (data saturation point). A similar approach was adopted for the left behind and vulnerable groups. Human rights advocacy organizations are few in Uganda, and the Human Rights Network (HURINET) was targeted for this information source category. The Uganda human rights commission was in particular targeted for issues regarding the HRWS and the LNOB groups.

A mix of data collection was employed. Primary data were collected from interviews of the key informants from government ministries, departments and agencies (MDAs); UN agencies; civil society organisations, comprising largely of NGOs (local and international NGOs); faith-based organisations; nomadic and indigenous communities; and the private sector. The majority of the key informants were interviewed on Skype/Zoom, while it was possible to organize one-on-one meetings with a few of them. Focus group discussions were held with representatives of left-behind groups, in informal settlements, mainly from Kampala. Secondary data were obtained from a review of unpublished and published documents, reports and journals at both the local/national and international levels. A review of the policies, laws and legislative frameworks was undertaken.

A variety of key informants were interviewed. Table 1 presents the category of key informants, and the specific organization or detail concerning each category of respondent. The informants are structured into three main categories. Category 1 comprises government MDAs. Category 2 comprises donor agencies, UN agencies, international organizations, NGOs, civil society agencies and private service providers. Category 3 comprises human rights advocacy organizations, LNOB and vulnerable groups.

Due to the prevailing lockdown situation caused by the COVID-19 pandemic, some of the interviews were done virtually. This affected the findings of the study, since the interaction that could have occurred in physical interviews was not realized during the online interactions. In the interviews which took place physically, more information in terms of literature/documents was obtained. During the online interviews, the interviewees often promised to send more information in the form of literature and documents and that information took a long time to be sent, and after several reminders, while other information was not sent at all. This work was largely qualitative. The quantitative data that were possible in this work were collected from the literature (documents and reports). No statistical analysis was adopted. For the qualitative data, content analysis was adopted.

RESULTS

The institutional framework for water and sanitation in Uganda has three levels: national, district and community. The national level comprises seven ministries, which are responsible for policy formulation, regulation, monitoring, planning and coordination, quality assurance and guidance, and capacity building. The ministries have several but joint liability to deliver water and sanitation. Uganda's MWE is principally responsible for setting national policies and standards, managing and regulating water resources, and determining priorities for water development and management. Working in conjunction with the MWE is the National Water and Sewerage Corporation (NWSC), a parastatal that is responsible for providing water and sewerage services to about 256 towns in urban centres across the country.

While the MWE bears the primary responsibility for delivering access to water and sanitation, it works together with seven other strategic ministries to realize its mandate, as follows:

i. the Ministry of Health (MoH), which is responsible for the promotion of hygiene and sanitation in households through its department of environmental health;
ii. the Ministry of Education and Sports (MoES), which is responsible for education in hygiene and the provision of sanitation facilities, as well as the handwashing after latrine use promotion in primary schools;
iii. the Ministry of Gender, Labour and Social Development (MoGLSD), which is responsible for gender responsiveness and community development/mobilization;
iv. the Ministry of Agriculture, Animal Industry and Fisheries (MAAIF), which is responsible for water use on farms and the management of water for farm production, including irrigation, animal husbandry and aquaculture;
v. the Ministry of Finance, Planning and Economic Development (MoFPED), which mobilizes and allocates funds to the different ministries;
vi. the Ministry of Local Government (MoLG), which is responsible for the creation, supervision and guidance of sustainable, efficient and effective service delivery in the decentralized system of governance; and
vii. the Ministry of Lands, Housing and Urban Development (MoLHUD), which is responsible for providing policy direction, national standards and coordination of all matters concerning lands, housing and urban development.

Table 1 | Categories of key informants and specific organizations or other details of respondents

No.	Category	Specific organization or other detail
1	Government MDAs	• MWE • MoGLSD • Uganda Human Rights Commission • KCCA
2	Donor agencies, UN agencies, international organizations, non-governmental organizations, civil society agencies and private service providers.	**UN agencies** • World Health Organization(WHO) • United Nations High Commission for Refugees(UNHCR) **Development partners** • Danish International Development Cooperation Agency (DANIDA) • United States Agency for International Development (USAID) **Civil society organizations** • Uganda Water and Sanitation NGO Network (UWASNET) • Human Rights Network (HURINET) **NGO** • Water For People • IRC (Country Director of IRC is country focal person for WSSCC) **Faith-based organization** • Diocese of Kigezi-Water Programme **Private sector** • Hydraulic and Sanitation Limited
3	LNOB, vulnerable groups	Elderly, women/girls, poor people in informal settlements, hospitals, Universal Secondary Schools, HIV/AIDS patients, Muslims at a mosque, sex workers; LGBTQ[a] could not be found in Uganda **Nomadic communities/indigenous communities** • Karenga District Local Government, Karamoja Region • Slum dwellers in Kampala

[a]In Uganda, the law does not recognize LGBTQ groups. The societal norms do not accept this and they are ostracized. Consequently, these groups could not be found in Uganda and were not included.

In addition, at the national level, within the sector-wide approach to water and sanitation, there is a Water Policy Committee and the Water and Sanitation Sector Working Group, mainly for facilitating the coordination of all activities on water and sanitation in the country. The members of the Water Policy Committee, established under the Water Statute of 1995, comprise representatives from government ministries, local government, the private sector and NGOs. The committee's basic mandate is to advise on water policy, standards for service delivery and priorities for water resources management. The Water and Sanitation Sector Working Group has the mandate of developing policy and then providing technical guidance.

At the district level, the institutional water and sanitation framework includes local governments empowered so they are responsible for appointing and managing private operators for urban piped water schemes that lie outside the jurisdiction of the NWSC. At the very tail-end at the community level, the community is responsible for demanding water supply and sanitation facilities. Water User Committees (WUCs) at the community level are responsible for mobilizing users to contribute to WASH governance. A summary of the national instruments and frameworks are presented in Table 2.

Development plans and regulations (local government)

The *Local Governments Act*[1] Section 31 provides for services to be provided at the request of a council. This gives authority to the Council (District Council, City Council, Municipal Council or Town Council) to give the people social services as delegated by the state in any sector concerned, including water and sanitation to ensure a suitable living environment without discrimination of any kind. To this end, the councils may pass local bylaws or ordinances, which apply only in their area of jurisdiction. These local laws must be aligned to the national laws and where there is conflict, the national laws prevail. A typical example of the local laws/ordinance is that of the Kampala Capital City on Sewage and Faecal Sludge Management. Local authorities can also prepare their strategies and plans aligned to the National Development Goals and aligned to the national annual budgets. An example of information relevant to the HRWS and the LNOB from the Kampala Sanitation Improvement and Financing Strategy is presented in Table 3 alongside information from the Kampala Sewerage and Faecal Sludge Management (FSM) Ordinance.

Regional and international instruments

SDG 6

The Government of Uganda, as a signatory to the UN SDGs, is working towards achieving SDG 6 by the year 2030. The MWE, with the support of UNICEF, has unpacked sector performance monitoring indicators (that included SDG 6.1 and 6.2 indicators) through a clear formulation of data collection tools and reporting templates. The indicators cover six themes (urban water supply, urban sanitation, rural water supply, rural sanitation, water for production and cross-cutting issues). In total, data are collected and reported on 42 indicators. Jointly with the MWE and MoH, GIZ developed a management information system to monitor safely managed/basic sanitation through the local government structures in urban areas. The tool was tested in municipalities in Northern Uganda. Along the way, some emerging issues were identified, requiring urgent attention if the sanitation sub-sector is going to achieve SDG 6.2:

i. Dissemination of the revised performance indicators for sanitation and hygiene to enable all stakeholders at different levels (planning, implementation, monitoring and data collection levels) to appreciate what is being measured and thus plan interventions accordingly.
ii. Migration from paper-based to computer-based management information system for sanitation and hygiene to achieve timely collection of data, ease analysis and ultimately improve the accuracy and reliability of the data.
iii. Menstrual hygiene management (MHM) has remained largely unattended to yet it is critical, especially to the retention of female children in school.
iv. Sanitation and hygiene at institutions like schools, healthcare facilities, prisons and barracks need urgent attention if the country is to attain universal access and 'LNOB'.

Mutual accountability mechanism

The government of the Republic of Uganda put in place the mutual accountability mechanism in 2010. Since then, the long-term National Development Strategy currently outlines several priority areas and strategies to improve the effectiveness of development assistance. Existing country mechanisms permit the joint assessment of progress in implementing some of the agreed Paris Agreement commitments on aid effectiveness. Additionally, donor-government forums, such as the Joint Budget Support Framework (JBSF), foster dialogue and mutual accountability.

DISCUSSION

A review was undertaken of the instruments, legislation/laws, policies, strategic and development plans and regulatory frameworks at the national, local government, regional and international instruments concerning the HRWS in Uganda. Development partners have also been active in promoting HRWS in Uganda through cooperating with the MWE, umbrella organizations under the ministry as well as NGOs to help in the supply of water and improving access to sanitation. This has been through funding programmes and projects of water and sanitation. The current programmes are being funded by a number of international bodies like the African Development Bank (AfDB), Austrian Development Agency and others in the Joint Water and Environment Sector Support Programme phase two (JWESSP II). This programme is under

[1] Act 1997 as amended.

Table 2 | National instruments and frameworks

No.	National instruments/frameworks	Provisions and analysis	Year
Legislation/laws			
1	Constitution of the Republic of Uganda	This is the Supreme legislation and has some articles that provide for the access to water and sanitation as human rights. Article 39 provides 'A right to a clean and healthy environment'. Articles 20 and 21 provides for the fundamental and other human rights and freedoms plus equality without discrimination.	1995 and amended in 2005 and 2018
2	The Water Act, Cap 152	Section 4(b) 'to promote a clean, safe and sufficient supply of water for domestic use for all purposes to people'. The whole of Section 7 is under the subtitle of 'General rights to use to water'. Section 16 'National water action plan' this section provides for the water action plan that is to be revised annually. Section 16 (1)(a) 'coordinate the preparation, revision and keeping up-to-date of a comprehensive action plan for the investigation, use, control, protection, management and administration of water for the nation'. Section 47 addresses the functions of water and sewerage authorities and the major function is provided in Section 47(1)(1)(a) 'to provide water supply services for domestic, stock, horticultural, industrial, commercial, recreational, environmental, and other beneficial uses as is required by the declaration establishing the authority or the performance contract'. Section 51 provides for water user associations that give powers for the operation of a water and sanitation committee in order to ensure communal supply of clean water leaving no person behind. Section 51(1) 'Where a water supply system is established by and serving more than one water user group, each operating through water and sanitation committee, the committees involved shall form a water user association which shall consist of an agreed representative of each committee.'	1997
3	Penal Code Act Cap 120	Section 176 provides for a punishment fouling water 'Any person who voluntarily corrupts or fouls the water of any public spring or reservoir, so as to render it less fit for purposes for which it is ordinarily used, commits a misdemeanour.'	2007
4	The Equal Opportunity Act	Section 14 provides for the functions of the equal opportunity commission. That is 'to monitor, evaluate and ensure that the policies, laws, plans, programs, activities, practices, traditions cultures, usages and customs of; organs of state at all levels, statutory bodies and agencies, public bodies and authorities, private businesses and enterprises, non-governmental organizations, and social and cultural communities, are compliant with equal opportunities and affirmative action in favour of groups marginalized on the basis of sex, race, colour, ethnic origin, tribe, creed, religion, social or economic standing, political opinion, disability, gender, age or any other reason created by history, tradition or custom'.	2007
5	Occupational Safety and Health Act	Section 51 (1) 'an employer shall provide and maintain for the use of workers, adequate and suitable washing facilities for washing, which shall be conveniently accessible and kept in a clean condition'.	2007

(*Continued.*)

Table 2 | Continued

No.	National instruments/frameworks	Provisions and analysis	Year
Policies			
6	Environment and Social Safeguards Policy	This policy is guided by 15 principles and 4 of these major principles uphold the ideas of LNOB to water and sanitation, human rights and gender equality in all the developments and projects. Principles 3, 4, 5 and 7 'Marginalized groups and vulnerable groups, human rights, Gender equality and women empowerment and indigenous groups' in order.	2018
7	NDP III	This is the third NDP for the period 2010/2021 to 2024/2025. Section 4.5.6 HRBA states that 'To ensure inclusive sustainable development, the NDP II has adopted the HRBA with particular attention to human rights principles of equality and non-discrimination, empowerment and participation and attention to vulnerable groups.' It states further, that 'Application of this approach will contribute to bringing the government closer to the people in order to effectively address their development needs, advance equality and leave no one behind. All sectors, ministries, departments, agencies and local governments are expected to adopt the HRBA in their respective policies, programs, legislation and plans.'	2020
8	Water and Environment Sector Budget	Parts C and K of the vote strategic objectives, on Page 1 of the National Water and Environment Sector Budget addresses 'Promotion of WASH in both rural and urban areas and promotion of gender equality'. Part C states 'To promote improved sanitation services in rural and urban areas including the promotion of handwashing with soap'. Part K addresses gender equality, stating 'To promote gender and equity considerations'.	2020
9	MWE Communication, Education, Participation and Awareness (CEPA)	The CEPA for the MWE is developed for the period 2019–2023. Section 1.1.1 provides for the vision of CEPA 'Sound management and sustainable utilization of water and environment resources for the betterment of the population of Uganda'. Section 1.3.2 gives the roles of CEPA, for example, 'Give voice to less powerful groups, minorities or disadvantaged groups' in the water and environment sector. This is very well aligned to the ideas of LNOB.	2019
10	Water and Environment Sector Strategic Investment Plan (SSIP)	The SSIP is for the period 2018–2030. It lists down 24 indicators and 7 of them are in line with LNOB, human rights and gender equality in water and sanitation. Indicator 1 is 'Village water supply'; Indicator 3 'Improved drinking water; Indicator 7 'Solid waste disposal'; Indicator 8 'Improved sanitation'; Indicator 9 'Safely managed sanitation'; Indicator 10 'Handwashing at home' and lastly Indicator 11 'handwashing at school'.	2018

implementation and will end in 2023. It is aiming at many projects. However, one of the core projects is to promote and provide clean safe water for all, thereby LNOB. It also focuses on gender balance in relation to WASH policies. The funding bodies are free to fund a project of their choice, for example, the AfDB is among the funders of community management.[2]

[2] JWESSP document.

Table 3 | Local ordinance and sanitation improvement and financing strategy for KCCA

No.	Instruments/frameworks	Provisions	Year
1	KCCA Sewage and FSM Ordinance	Section 6(2) 'The minimum standards for onsite sanitation technology' options shall require that every onsite sanitation technology: (b) 'has a handwashing facility with detergent' (e) 'allows for inclusiveness including provisions for persons with disabilities' (f) 'provides for menstrual hygiene management' (j) 'is kept clean and hygienic'. These provisions are in line with the LNOB ideas.	2019
2	Kampala Sanitation Improvement and Financing Strategy	Section 1.2 states the purpose of this strategy as 'achieving target 6.2 of the SDG that is, achieving safely managed sanitation for all'. Section 1.2 states that 'The strategy aims to achieve specifically Sustainable Development Goals that is achieving safely managed sanitation in homes, schools, health centers and public places.' 'Gender equality, poverty and other inequality are also to be addressed.' Section 2.1 of the Strategy addresses the objectives of this Strategy of which two of them relate to LNOB ideas. Objective 1 'To achieve equitable and universal access to improved sanitation and hygiene and safe management of faecal sludge (FS) and sewerage along the entire sanitation chain in Kampala by the year 2030,' where in "Equitable implies progressive reduction and elimination of inequalities between population subgroups and universal implies facilities close to homes that can be easily reached and used when needed". Section 2.2 addresses the principles of the strategy. Principles 1 and 2 are outstanding in regard to LNOB. That is 'Systems approach to sanitation' and 'inclusiveness'. These principles advocate for 'Development of approaches and measures to reach the most vulnerable, especially focussing on urban poor, gender and the disabled'.	2020

Regarding equity in accessing WASH services, there is a Presidential Directive to ensure that there is a water source per village, recognized by a Local Council I, irrespective of the size or population of the village in Uganda. This has been put in place to ensure that water supplies reach every village, thereby LNOB.

In the capital and other cities, municipalities and towns, there are deliberate efforts to reach out to everyone in terms of water supply and sanitation. The NWSC, a water and sewerage utility, 100% owned by the government of the Republic of Uganda, has pro-poor strategies of connecting low-income communities and vulnerable groups with a subsidized tariff structure. In so doing, they eliminate the middle-men, for instance water vendors, by installing pre-paid water meters in the slum areas. This is widespread in informal settlements in Kampala, mainly in Ndeeba, Bwaise, Kinawataka, Kisenyi, Banda, Kibuye and Kamwokya, to mention but a few, but it is not yet rolled out to other cities, municipalities and towns. The MWE embarked on the pro-poor policy in 2006 and the NWSC started implementing it in the same year. It is being rolled to all 256 towns where the NWSC operated. In rural areas, it is implemented by the MWE, and in small towns, by the Water and Sanitation Development Facility.

The local authorities, for instance Kampala Capital City Authority (KCCA), are implementing an FSM programme which promotes the construction of safely managed sanitation facilities in low-income households, public schools, public health centres and markets, targeting vulnerable and left-behind groups. In these areas, besides the construction of safely managed sanitation facilities, KCCA dislodges latrines and septic tanks serving left-behind and vulnerable groups for free. KCCA operates one faecal sludge truck for each of the five divisions that make up Kampala. KCCA works with the private sector and the NWSC to ensure that the coverage of water supply and sanitation is expanded in order to reduce/limit travel distances to access these critical services. NWSC hires private companies in the water supply network to left-behind and vulnerable groups. The private sector is mainly operating in the collection and transport of faecal sludge at a fee regulated by KCCA. In Kampala, there are more than 100 trucks owned and operated by the private sector, which empty faecal sludge on a daily basis and transport it for treatment to the NWSC operated treatment plants. KCCA together with the National Environment Management Authority (NEMA) is responsible for licensing and regulating the truck operators dealing in the business of collection and transportation of faecal sludge. While the other cities, municipalities and towns do not have a faecal sludge management programme, they do have the private sectors who build sanitation facilities and those that collect and transport faecal sludge at a fee. In towns other than Kampala, there is no involvement of the local authority (city/municipality/town council authorities) in regulating prices. In some cases, these are exorbitantly high and therefore prohibit residents hiring these services.

MHM is clearly a left-out issue, especially among the LNOB groups. Nearly all responses from all groups fetched no, or to a limited extent, little data pertaining to ongoing MHM initiatives. There is therefore a need to identify and put in place measures for the management of MHM products. While incinerators are being promoted in schools and health centres for the destruction of MHM products, there are cultural taboos related to the burning of MHM products. In some communities in Uganda, especially in the eastern part of the country (Busoga sub-region), incinerating menstrual pads is considered taboo, as it is tantamount to incinerating 'one's soul'. It is therefore important to involve the women and girls in identifying and determining how their MHM products should be disposed of. Socio-cultural myths should be clearly understood and unpacked prior to investing in sanitation facilities. Community leaders should be involved in mobilizing the community members to participate in home and village improvement campaigns as well as a campaign at the schools and other institutions to improve upon hygiene.

More so, existing strategies for sanitation provision leave out some groups and/or situations, for instance public – roadside sanitation – markets. Besides, there is a need to determine the access levels, in terms of safely managed sanitation for communities in informal settlements; institutions (schools, healthcare facilities, offices, hotels and prisons) and households in both rural and urban areas.

There are some efforts in place to operationalize the LNOB principle in water governance. For instance, the Third National Development Plan (NDP III) (National Planning Authority 2020) recognizes the human rights-based approach (HRBA) as a way of bringing the government closer to the people in order to effectively address their development needs, advance equality and LNOB. Accordingly, the NDP III calls upon all sectors, MDAs and local governments to adopt the HRBA in their respective policies, programmes, legislation and plans.

In terms of budgeting and financing, the government ensures community participation in planning, budgeting and service provision. There exists an administrative structure, which ensures the collection of information right from the village, through the sub-national level to the national level. In Uganda, local governments feed into the budgeting process (MoFPED 2009). The budgeting starts with the issuance of the budget framework by the MoFPED, which gives ceilings for each accounting unit. Although the accounting units are districts and municipal local governments, there are also lower-level local governments, that is, the sub-counties, which are accountable to the districts and the districts are accountable to the national government. A budget framework paper is presented at the district/municipal level. The elected representatives of the district or municipal council, who include representatives from each sub-county, reveal the allocations to their sub-counties and request input from village leaders on what they need to be included/prioritized in the budget. The village council leadership prepares a meeting of the village members to give input into their priorities to be included in the budget. Communities contribute to their WASH projects, in terms of co-financing or in-kind for those who cannot afford it. The MWE started reporting the HRWS in 2018.

The reporting on HRWS is embedded in the 42 indicators, which the Water and Environment Sector is mandated to report upon annually. The HRWS- and LNOB-related indicators include, but are not limited to: access to safely managed water in terms of the percentage of population using safely managed drinking water services located on premises; the percentage of villages with a source of safe water supply; access to safely managed sanitation in terms of the percentage of population using safely managed sanitation services; and gender in terms of the percentage of WUCs/water boards/environmental management/water catchment management committees (CMCs) with women holding key positions. The budgets on the HRWS are integrated into the entire Ministry's budget. According to the information provided by the MWE, there is no single budget line allocated to HRWS, because the budget that addresses HRWS is integrated into many indicators, which are elucidated above.

As far as a quality management system (QMS) is concerned, the MWE has a QMS integrated into the Sector Performance Measurement Framework. However, information on this could not be easily found from any source. The most recent Water and Environment Sector Performance Report of 2019 mentions a water QMS. In relation to the QMS, the WESPR (Ministry of Water and Environment 2019) indicated that the government is continuing to implement the National Water Quality Management Strategy through the upgrading of the Entebbe water quality laboratory to a national reference laboratory, the establishment and operation of regional laboratories in the water management zones (WMZs) and the development of water quality guidelines and standards for various emerging issues such as oil drilling and emergency response.

In terms of capacity, this study established that mindset change is promoted and accomplished through awareness-raising. The awareness is being raised on the gender mainstreaming guidelines, performance measurement framework, and vulnerable groups. All unserved communities are considered vulnerable groups. The capacity of women is being built to become members of the water users' association and CMCs. According to the gender strategy of the MWE, the target is to reach a membership of 50% women. Currently, 35% of CMCs have women holding key positions (Ministry of Water and Environment, 2019). In the financial year (FY) 2018/2019, it was established that 75% of the water user management committees had women in key positions and 49% of water dams had women in key positions (Chairperson, Vice-Chairperson, Secretary and Treasurer). In the WMZs, out of the nine sub-CMCs formed, 78% have women occupying key positions. The SMCs without women in key positions include Opeta Bisina and Agago.

In the FY 2018/2019, the MWE in conjunction with the Nile Basin Initiative Secretariat coordinated and participated in the training of 30 decision-makers (7 women and 23 men) in Transboundary Water Cooperation, Hydrodiplomacy and International Water Law. Participants were drawn from the Sector and other line ministries and agencies of: Foreign Affairs, Internal Affairs, Defence and Veteran Affairs, East African Community Affairs, Finance, Planning and Economic Development, Works and Transport, Local Government, and Energy and Mineral Development. This helped in the building of the national capacities for the coordination of the transboundary water resources activities. Furthermore, in the FY 2018/2019, capacity building and training in gender-inclusive integrated water resources management and effective catchment management were conducted with a total of 3,301 participants (38% being women).

Regarding gender analysis in terms of staffing in the MWE, the senior management position indicates that there are 38 staff members in management positions, out of which 16% (6) are female. At the middle management level, there are 123 staff members, of which 26% (31) are female and 74% (90) male. The percentage of females in this category has increased from 23% reported in FY 2017/2018 to 26% in FY 2018/2019. At the operational level, women are fairly represented with females constituting 46% out of the 85 staff under this category.

As for monitoring and evaluation, the sector performance management framework is in place; and every year, there is the water and environment review. Since 2005, a WESPR is produced by September every year. Data are collected, although not in a disaggregated way. Since 2018, UNICEF supported the MWE to report on SDG 6. Data are collected on 42 indicators, covering SDG 6 and they include HRWS and LNOB groups. Real-time data collection is problematic and non-existent. Existing data collection systems are largely manual, do not collect and update real-time data and are largely inefficient. Quite often, the village health teams who collect data at the village level lack facilitation in terms of physical transportation (car, motorcycle or bicycle) to transmit data for management, to sub-county, for onward transmission to the district or the national level, wherever it is needed. They lack modern methods of data collection and transmission, for instance by mobile phones. Large-scale water supply systems in Kampala are equipped with supervisory control and data acquisition (SCADA), which enable the collection of real-time data on the water treatment plant operational parameters including, for instance, water treated and pumped into distribution. Such systems are lacking in small water supply systems and have not been deployed to collect information on sanitation and hygiene.

In terms of tariff subsidies, according to a study by the International Bank for Reconstruction and Development/the World Bank of 2015, Uganda has been remarkably effective in delivering services to the poor and economically disadvantaged people. However, the fiscal and political context that facilitated this success changed, and the bias against rural water services seems likely to continue as a result. This was found to disproportionally harm poor and economically disadvantaged people, as they live overwhelmingly in rural areas (IBRD/WB 2015). The study by the World Bank recommended that the pro-poor strategy be overhauled to put in place measurable objectives and outcomes, and a monitoring system to hold the Directorate of Water Development (DWD) accountable for implementation; and align these outcomes to the SDGs. The second recommendation comprised a set of suggestions to the content of the new strategy, to target: (1) reducing piped scheme tariff subsidies; (2) assisting local water authorities and local private operators and individuals to integrate the management of hand-pumped supplies and piped schemes and (3) developing targeted programmes to improve access and other aspects of improved water services to both the poor and the bottom 40%. The pro-poor strategy has not been updated or overhauled in line with the IBRD/World Bank Study, and hence the study recommendations have not been implemented.

Action number eight in the pro-poor strategy calls for empowering communities through participation to ensure cross-subsidy in cost recovery, arguing that 'Communities themselves are in the best position to define who is poor and the most appropriate way to assist them (e.g., support in-kind and cross subsidy for capital and O&M contributions). If the communities are empowered by being involved in key decisions and trained to undertake the main responsibilities, the facilities will be better managed and will be more effective in serving the poor.' The IBD/World Bank study in 2015 found that community participation was lower than hoped, despite participatory procedures put in place by the DWD. In a survey commissioned by DWD (2009), which looked at community participation throughout the project cycle for 160 water supplies, unexpectedly low levels of participation, particularly in respect to the choice of technology, were encountered. That could mean that community members wished for a more expensive level of service, but instead received a less costly one, for example, a borehole instead of a piped scheme, or a protected source instead of a borehole. Furthermore, the DWD survey (2009) provided some evidence of the 'cross-subsidy' called for in Action number eight, although not to the extent intended. Interviews with water committee members did reveal that certain vulnerable groups were exempted from operation and maintenance contributions, but poverty was generally not a reason for exemption, unless it was combined with elderliness or disabilities. Only 15% of respondents said that poor people were subsidized by virtue of being poor.

USAID is embarking on a programme to support the Water and Sanitation Umbrella organizations to reduce water connection fees for poor households. USAID is supporting the construction of sanitation facilities in schools, using the MoH guidelines to cater for access by the persons with disabilities. USAID is supporting the institutional strengthening using the scorecard adopted by the Government of Uganda, to assess the sanitation situation in institutions.

The private sector is also not left behind in supporting inclusive WASH services. WASH equipment, for instance the rainwater harvesting equipment, is readily available on the market, the high costs notwithstanding. The private sector is also actively engaged in innovation to develop appropriate WASH solutions such as compact mini water treatment suitable for small users such as a hotel. This is especially for hotels and schools located in areas off-grid, where no existing piped water supply systems are in place. The collection and transportation of faecal sludge to treatment plants is private sector-led. There are more than 100 vacuum trucks belonging to the private sector, which collect and transport faecal sludge in Kampala on a daily basis. Each of these trucks is able to make three to four trips per day, thereby enabling them to make a profit of at least UGX 150,000–200,000 per day. Some towns within Uganda lack vacuum trucks. Consequently, customers needing to empty their sanitation containment systems (pits and septic tanks) are outsourced from other towns and the distances sometimes exceed 100 km one way, making it a 200 km return journey. This makes the collection and transport of faecal sludge extremely expensive. Small-scale private emptiers using semi-mechanized emptying equipment such as gulpers can fill in the gap, where mechanized/vacuum trucks are not available. However, small-scale emptying equipment is limited in terms of how far the accompanying transportation may reach. Unfortunately, a number of towns do not have faecal sludge treatment plants, necessitating the transfer of sludge from one town to another.

DISCUSSION AND CONCLUSION

While Uganda is a signatory to the UN resolution that seeks to achieve the UDHR including the right to water and sanitation, there are still many categories of society who are glaringly left behind in access to water and sanitation. There exists a wide inequality in water access service levels between rural and urban areas of Uganda (WHO, 2019). There also exist disparities in access between different interest groups. Categories of left-behind groups or vulnerable populations include the elderly, women, children and students, nomadic populations, extremely poor populations, refugees and internally displaced people, people with physical or developmental disabilities, geographically dispersed or hard-to-reach populations and groups excluded based on sexual orientation such as LGBTQ. Despite the glaring gaps in universal water access, the first line of defence in the prevention strategies for the current COVID-19 pandemic is frequent handwashing with soap. Therefore, there is a need to deliberately put in place efforts pursuing the right to access. At the level of the MWE, and the NWSC, this should have measurable indicators, for instance how many people to serve per year, and there should be incentives for meeting the intended targets and disincentives for failure to achieve the targets.

The challenge is, and has always been, in implementation. One of the key bottlenecks identified virtually by all key informants is the issue of inadequate funding for the sector. Critical issues for the Water and Environment Sector still include inadequate financing, capacity gaps and lack of subsidy arrangement to support the acquisition of sanitation facilities as well as adequate and safely managed water supplies for the poorest of the poor. Inadequate financing of the Water and Environment Sector is considered a major challenge hindering the fulfilment of their functions. As a result, the targets under the Strategic Sector Investment Plan (2018–2030) and Presidential Directives (e.g. one water source per village) are unlikely to be met. Capacity gaps in the sector are critical issues, particularly in newly created local governments. Furthermore, subsidies for household sanitation are seen to be critical for attaining full sanitation coverage. However, for the past two decades, the government has maintained the policy of no subsidy for household sanitation. Although this policy has had varying success with some areas attaining high basic sanitation coverage, others have remained behind, especially in areas with extremely poor households such as informal settlements.

While the Government of Uganda has made tremendous efforts in pushing for access improvement, enormous challenges and gaps remain. For instance, the pro-poor strategies to access safe WASH facilities are largely insignificant, as they do not address tariff reduction measures; yet, the tariff is still by and large prohibitive. Secondly, the private sector decried limited support from the government first, in providing adequate incentives for doing business, and secondly in providing the required market. Instead, government reportedly prefers imported technologies over and above local content. Thirdly, there are also concerns about efficiency gains. While prior to 2005, government investments in the water sector were guided by Strategic Investment Plans which were guided by real need and demand. However, this approach had later to be abandoned, as it was not politically appealing. The government investments in WASH since 2001 are now largely directed by political decisions. Nobody seems to be asking questions of expenditure per capita in WASH investments to underscore efficiency.

It should be noted that the existing tariff structure is not favourable to the poorest. The cross-subsidy is small and has a very low trickledown effect on the poor unless they are elderly or have disabilities. While on paper the HRWS features prominently, in reality, water is considered an economic good where users must pay for cost recovery.

Governance challenges of the WASH facilities remain enormous. Sustainable O&M is still elusive. For instance, in Amuru district in northern Uganda, a recent audit supported by USAID found that 34% of the water sources were no longer functional. Moreover, when it comes to sanitation, the challenges and gaps are even more pronounced. Water and sanitation interventions are normally combined within the same projects, with the water component taking the lion's share.

While WASH-sector players easily produce data on water supply, this is not the case for sanitation and hygiene because these are mainly practised at the household level and data are captured mainly through surveys. Moreover, research is mainly promoted by non-state actors. The government ministries hardly allocate funds to do research on WASH issues. The collection of real-time disaggregated data is not happening. There is a need for adequate capacity (both human), and equipment to undertake real-time collection of disaggregated data as well as more investment in research to inform policy and practice, as well as to collect disaggregated data. There is a need to promote local content and use locally available technologies such as simplified SCADA systems to enhance real-time data acquisition.

The HRWS and the LNOB are not yet known by many practitioners within the sector. Moreover, even for those who know it, it is not clear what these embed. The available strategies have very few statements on gender and HIV/AIDS. The ministry of education lacks an MHM strategy. There exist sufficient policies, laws and legislation and enforcement frameworks, which should be updated in the context of the HRWS and the LNOB. Most existing laws, policies and legislation were promulgated before or during the predecessor Millennium Development Goal time. SDGs are more stringent and therefore, the laws, policies and legislative frameworks to deliver them, should be aligned. Community dialogue to create sensitization, awareness and understanding of the HRWS is therefore critical. Targeted advocacy and sensitization should now focus on the politicians, since they are the decision-makers in the allocation of resources both at national and district levels such as the Members of Parliament and the district councillors.

Climate change is an emerging concern that influences the design and functioning of WASH infrastructure. In hard-to-reach areas in mountains in north-eastern and western Uganda, flooding separates villages and interrupts water supply. Some intake works of gravity flow schemes are totally destroyed, resulting in lack of supply service for many months. There is a concern that the rising groundwater levels due to climate change limit the vertical separation distance between groundwater levels and pit latrines. Local governments, with the support of the MWE, should zone their areas of jurisdiction, with guidance on how deep sanitation containment systems should be sunk. In general terms, the current environmental conditions are characterized by population pressure forcing settlement in marginal areas, both in the rural and urban areas. In the rural areas, population pressure manifests in overgrazing, severe soil erosion and soil exhaustion. In urban areas, this manifests in the occupation of wetlands and the proliferation of slums/informal settlements in these marginal areas. Compounded with the climate change-related effects such as periodic floods and droughts, in the different parts of the country, the impacts on the bio-physical environment have brought a lot of suffering to the Ugandan population living in rural and urban areas. Moreover, in hard-to-reach areas, such as mountainous areas, the technologies for the supply of WASH services become expensive, leaving them behind. In this regard, climate change needs to be embedded in existing documents and, most importantly, to fund their implementation, including enforcement.

DATA AVAILABILITY STATEMENT

All relevant data are included in the paper or its Supplementary Information.

REFERENCES

DWD 2009 *Mapping a Healthier Future: How Spatial Analysis Can Guide Pro-poor Water and Sanitation Programs in Uganda.* Water Resources Institute, Washington, D.C./Kampala.

International Bank for Reconstruction and Development/The World Bank, IBRD/WB 2015 *Republic of Uganda. Water and Sanitation for the Poor and Bottom 40% in Uganda: A Review of Strategy and Practice Since 2006.* Report No. ACS13381.

Ministry of Water and Environment (MWE) 2019 *Water and Environment Sector Performance Report 2019.* Ministry of Water and Environment, Government of Uganda, Kampala. https://www.mwe.go.ug/sites/default/files/library/SPR%20FINAL%20BOOK%202019.pdf. (accessed 10 August 2020).

MoFPED 2009 *A Guide to the Budget Process.* Ensuring that Uganda's National budget reflects the views and priorities of citizens. Available from: https://budget.go.ug/ (accessed on 25 September 2020).

Olsen, C. & Marie, D. M. 2004 *Cross-sectional Study Design and Data Analysis.* College Entrance Examination Board, New York, USA.

Pahl-Wostl, C. 1995 *The Dynamic Nature of Ecosystems Chaos and Order Entwined.* John Wiley & Sons, Hoboken, NJ.

National Planning Authority 2020 Uganda National Development Plan III. Available from: https://www.npa.go.ug/wp-content/uploads/2020/08/NDPIII-Finale_Compressed.pdf (accessed 27 September 2020).

Universal Declaration of Human Rights. Available from: https://www.ohchr.org/EN/UDHR/Documents/UDHR_Translations/eng.pdf (accessed 24 August 2020).

World Health Organization 2019 Trends in maternal mortality 2000 to 2017: estimates by WHO, UNICEF, UNFPA, World Bank Group and the United Nations Population Division. World Health Organization, Geneva, Switzerland.

First received 25 February 2021; accepted in revised form 11 August 2021. Available online 30 August 2021

IWA PUBLISHING

 Water Management Addressing Societal and Climate Change Challenges doi: 10.2166/wcc.2021.190

Future virtual water flows under climate and population change scenarios: focusing on its determinants

Niloofar Ashktorab * and Mansour Zibaei
Department of Agricultural Economics, School of Agriculture, Shiraz University, Shiraz, Iran
*Corresponding author. E-mails: ashktorab@asnrukh.ac.ir; nilo.ashktorab@gmail.com

NA, 0000-0002-2855-879X

ABSTRACT

Water resources are under increasing pressure from population growth, socio-economic development, and climate change. The main user of freshwater is the agriculture sector, accounting for 70–80% of global water use. Consequently, increasing water productivity and reducing water use in this sector are vital for alleviating water scarcity. It has been alleged that domestic or international trade of water-consumed products from wet-provinces or countries to arid and semi-arid provinces or countries is one possible path to mitigate water shortage. In this regard, virtual water flows and their determinants in Iran are investigated in this paper. This study examines the main determinants of bilateral virtual water flows associated with international trade in agricultural goods among Iran and other countries using a gravity model of trade. The impacts of climate change and socio-economic change scenarios on the virtual water trade are simulated. The result of the gravity model indicated that bilateral virtual water trade flows are influenced by the classical determinants and all the mass-related variables (gross domestic product and population) are significant with expected signs. Finally, the result obtained from the simulation of climate and socio-economic change scenarios showed that under the pessimistic scenario of climate change and population growth rate of 0.98 and 0.44%, VW trade between Iran and other countries would increase by 41 and 66% in the periods of 2016–2045 and 2070–2099, respectively. As a consequence, identifying the appropriate countries for business interactions and adopting effective trade policies are very important and need attention soon.

Key words: climate and demographic change scenarios, determinants, Iran, virtual water trade, water footprint

HIGHLIGHTS

- Bilateral virtual water trade flows are influenced by the classical determinants and all the mass-related variables (gross domestic product and population).
- The size of the future virtual water flows was predicted under the climate and population change scenarios.
- The average annual temperature and the population of Iran, as the determinants of virtual water flows, have a positive and significant effect on the net virtual water flows.

GRAPHICAL ABSTRACT

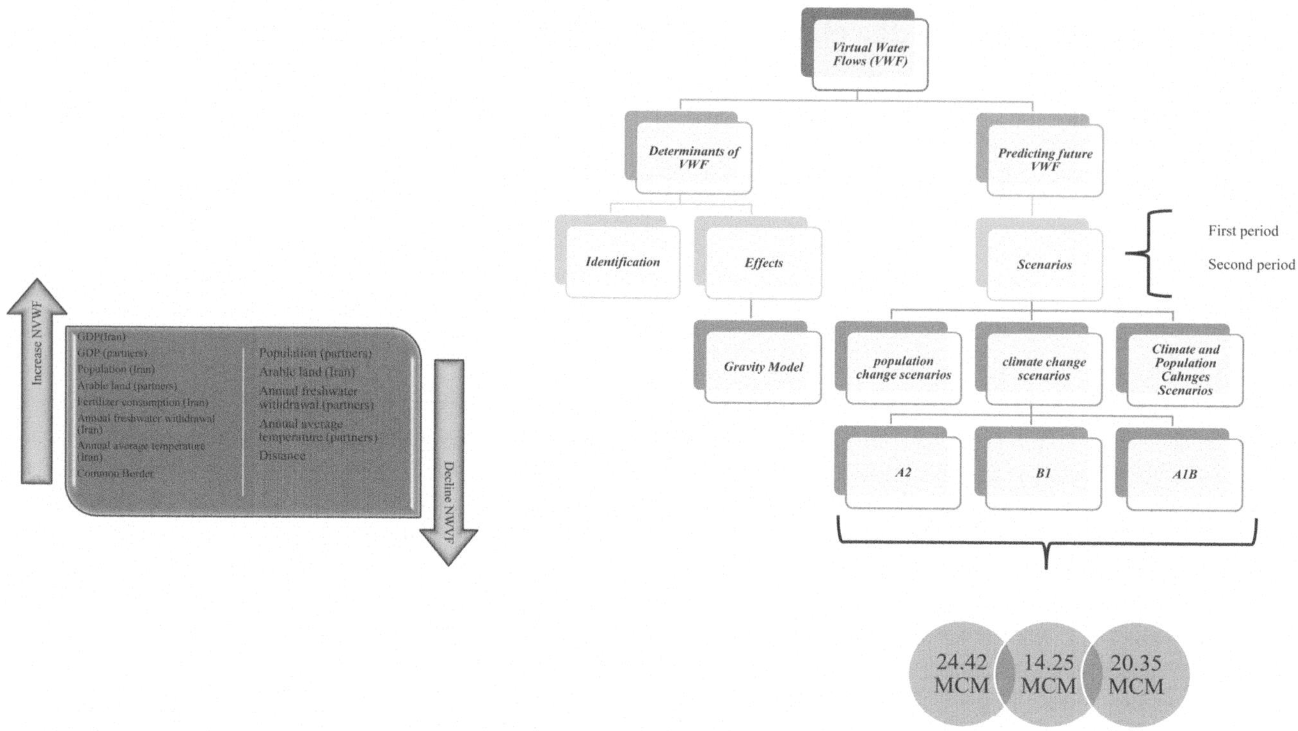

INTRODUCTION

Factors such as socio-economic development, population growth, and climate change have created major challenges for sustainability and food security as two of the most important societal concerns (Vorosmarty *et al.* 2000; Gleik 2008). In this regard, the water resources, which are essential for human life, sustainable livelihoods, and ecosystems, are under pressure. While certain areas of the world have abundant water, other areas suffer from water shortages. Iran is considered as one of the areas facing severe water scarcity. The water *per capita* in Iran was 5,570 cubic meters in 1962, which declined sharply to 1,876 cubic meters in 2008 and 1,688 cubic meters in 2017, which is lower than the world average (6,225 cubic meters) (Aquastat 2017; World Bank 2014). Various factors such as climate change and diversity, rapid economic development, widespread urbanization and pollution, and irrational agricultural policies may justify unprecedented pressure on water resources (FAO 2016). The current status of water resources in Iran and its governing trends and processes clearly illustrate the need for managing the demand and moderating water consumption. The agricultural sector, with the highest water consumption (80% of the total consumption; Rost *et al.* 2008), has focused on reducing water consumption. The improvement of water productivity is considered as the most important strategy to balance the declining supply and increasing demand for water resources. There are several ways to improve the productivity and increase water use efficiency. In the same vein, the international trade of the agricultural products is a way to improve the global water use efficiency through the virtual transfer of water resources to the water-stressed areas (Dalin 2014), which saves a significant amount of water resources globally or nationally (Chapagain *et al.* 2006). The water used in production processes is called virtual water, which is transferred in the form of trading the agricultural, livestock, and industrial products and is used as a strategic tool to increase water access in areas facing water constraints (Mubako 2011).

In the case of the products containing virtual water, trade is regarded as a means for transferring water resources between regions. In addition, the domestic and international food trade saves water on a national scale by encouraging the virtual water exchanges from provinces or countries with higher production productivity to the regions with less production productivity and reduces the water consumption in producing per unit of production (Chapagain *et al.* 2006). Trading the water-consumed products reduces the spatial heterogeneity of the access to water resources. Similarly, the domestic or international trade improves the water use efficiency at the global or national scale by providing efficient products for all consumers (Dalin 2014).

Since the efficiency of agricultural water use, depending on the type of technology and biophysical factors, varies widely in different regions, the international or domestic food trade saves a significant amount of water resources globally or nationally if the regions with higher water resource efficiency export water to the less efficient regions (Chapagain *et al.* 2006). Measuring the international and domestic flow of water resources through the food trade contributes to the identification of the key exporters and importers and the performance evaluation of the trading system about the global water use (Dalin 2014).

The virtual water flows indicate the amount of virtual water exports and imports which are affected by various factors. Some of these factors such as population and gross domestic product (GDP) are considered in the country size group, and some other factors such as temperature, arable lands, and water resources inventory as natural resources affect the virtual water flows. The third group, known as the economic group, includes variables such as tariffs on agricultural products in the destination country, number of tractors in the destination country, amount of fertilizer consumed per hectare in the country of origin, and price of irrigation water. Each group of these variables affects the virtual water flows directly or indirectly (Fracasso *et al.* 2016). The predictions indicate that future food demand and supply change with the joint or independent impact of the climate change and population. Climate change limits global access to water and food production while increasing water consumption in all economic sectors (DaMatta *et al.* 2010; Strzepek & Boehlert 2010; Yawson *et al.* 2020). Furthermore, the population change directly and indirectly influences the global food and water supply and demand (Huang *et al.* 2010; Foresight 2011; Thomson *et al.* 2013).

The importance of virtual water flows has led to many studies in this area (Table 1), some of which have focused on quantifying the issue globally (Hoekstra & Hung 2002; Zimmer & Renault 2003; Islam *et al.* 2007; Lenzen *et al.* 2013; Biewald *et al.* 2014; Dalin 2014; Ercin & Hoekstra 2014; Verhofstadt *et al.* 2016), and of which some have examined the issue at the specific region (Wang *et al.* 2013; Zoumides *et al.* 2014; Casolani *et al.* 2016; Deng *et al.* 2016; Zhuo *et al.* 2016; Chouchane *et al.* 2018; Zhang *et al.* 2018). Factors affecting the virtual water flows and the future of the virtual water flows under different scenarios have been investigated in several studies (Fader *et al.* 2010; Pfister *et al.* 2011; Hanasaki *et al.* 2013; Konar *et al.* 2013; Liu *et al.* 2013; Ercin & Hoekstra 2014; Haddeland *et al.* 2014; Wada & Bierkens 2014; Zhu *et al.* 2015). Other studies examine the virtual water trade from an international perspective, considering water resources as well as water scarcity (Oki & Kanae 2004; Islam *et al.* 2007).

Generally, the situation is likely to become worse in the near future and it is reasonable to anticipate more competition for the water resources (United Nations 2015). In addition, based on the estimation on the national income growth provided by the World Bank (World Bank 2016) and the population growth by the World Population Prospect (United Nations 2019), it is predicted that the water demand for industrial activities and urban needs increases in the future. The significant increase in the water demand over the renewable domestic water resources, water scarcity, and its implications for food security have created major concerns in most of the countries (Ashktorab 2019).

Since Iran is heavily under water stress, it is necessary to consider all strategies mitigating the pressures on water resources by increasing water productivity and managing its demand. The present study examined the virtual water trading and adopted a clear strategy to use all the opportunities created and avoid the threats resulting from its application. Furthermore, it aimed to identify the determinants of bilateral virtual water flows associated with the international trade of the selected agricultural products such as wheat, barley, maize, rice, sugar beet, and tomato, which allocated 72% of the arable lands in Iran (Ashktorab 2019), and predict the size of the virtual water flows of the selected agricultural crops influenced by climate change and population scenarios in the future.

MATERIALS AND METHODS

Determinants of the virtual water flows in Iran

The idea of virtual water flows was originally proposed by Allan (1993, 1998) based on the economic concepts developed in the international trade literature, particularly in the Heckscher–Ohlin Vanek model (Fracasso *et al.* 2016). The results of numerous studies indicated that there are many variables playing a significant role in virtual water flows (Figure 1), which are divided into four groups, some of which are evaluated in the regional or global studies (Yang *et al.* 2003; De Fraiture *et al.* 2004; Ramirez-Vallejo & Rogers 2004; Yang & Zehnder 2007; Novo *et al.* 2009; Lenzen *et al.* 2012; Debaere 2014; Fracasso 2014).

Table 1 | Virtual water flows studies along with their ability to solve challenges

			Challenges				
NO.	Title	Year	Size of virtual water flows	Determinants of virtual water flows	Climate change scenarios	Gravity model	Predicting the future
1	Virtual water flow in food trade system of two west African cities	2019	✓	✗	✗	✗	✗
2	Drivers of virtual water flows on regional water scarcity in China	2019	✗	✓	✗	✗	✗
3	Long term drivers of global virtual water trade: a trade gravity approach for 1965–2010	2019	✗	✓	✗	✓	✗
4	Water footprint of wheat under climate change: trends and uncertainties associated to the ensemble of crop model	2019	✗	✗	✗	✗	✓
5	Global implication of regional grain production through virtual water trade	2019	✓	✗	✗	✗	✓
6	Influential factors on water footprint: a focus on wheat production and consumption in virtual water import and export region	2019	✗	✓	✗	✗	✗
7	Evaluation of agricultural water demand under future climate change scenarios in the Loess Plateau of northern Shaanxi, China	2018	✗	✗	✓	✗	✓
8	Economic aspects of virtual water trade (Oki *et al.* 2017)	2017	✓	✓	✗	✗	✗
9	Determinants of virtual water flows in the Mediterranean	2016	✗	✓	✗	✓	✗
10	Modeling the future evolution of the virtual water trade network: A combination of network and gravity models	2016	✓	✗	✗	✓	✓
11	Inter and intra-annual variation of water footprint of crops and blue water scarcity in the Yellow River Basin (1961–2009)	2016	✓	✓	✓	✗	✓
12	Balancing water resource conservation and food security in China	2015	✓	✓	✓	✗	✓
13	The global economics of water: Is water a source of comparative advantage?	2014	✗	✓	✗	✓	✗
14	Water footprint scenarios for 2050: a global analysis	2014	✗	✓	✓	✗	✓

The population and GDP of the exporting and importing countries are situated in the first group, and the population is expected to have a positive impact on the virtual water flows of the countries. The second group is related to the water inventory. Many studies indicated that the countries with inadequate water resources are net exporters of the virtual water (De Fraiture *et al.* 2004; Ramirez-Vallejo & Rogers 2004; Yang & Zehnder 2007; Verma *et al.* 2009; Roson & Sartori 2010, 2013). Among the variables related to the water inventory, four variables are considered for both exporting and importing countries. The first is the annual freshwater withdrawals for agricultural purposes. The second is related to the total green water and the third variable is called the water availability ratio, which examines the pressure on the water resources and is considered as the ratio between the total available water and the water needed for the food pattern of each country. Finally, the average annual temperature is considered as the fourth variable (Fracasso *et al.* 2016).

The amount of the available water cannot be considered as the main and only variable affecting the virtual water flows. Kumar & Singh (2005) demonstrated that the water resources should be considered in relation to the arable land. Therefore, the agricultural lands of the exporting and importing countries are both considered as the factors influencing the virtual water flows (Fracasso *et al.* 2016). Some factors, such as production technologies, domestic and international prices of goods, trade barriers, and other economic and social characteristics, can affect the production, consumption, exchange of agricultural

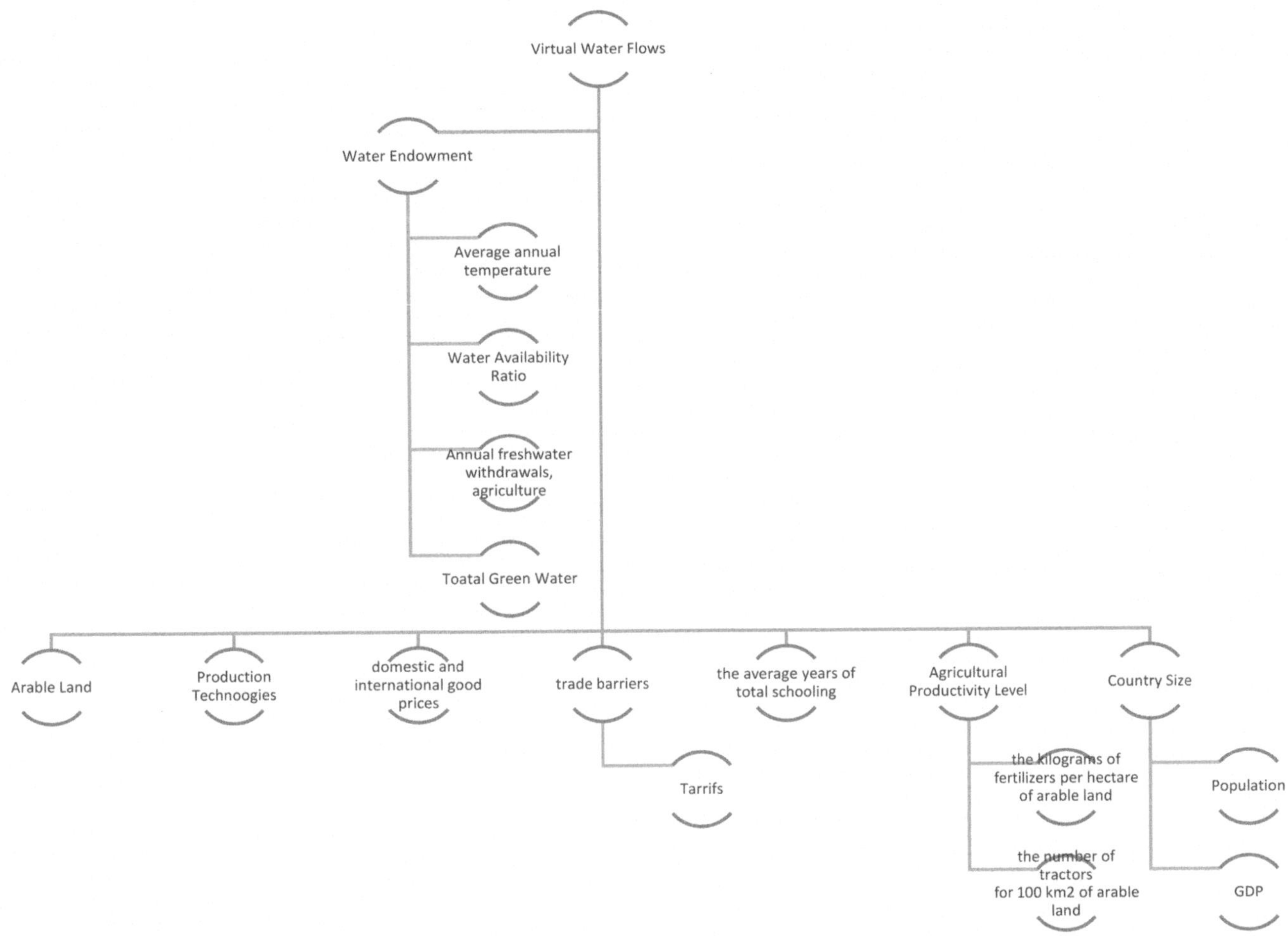

Figure 1 | Determinants of virtual water flows.

goods, and the water available in the agricultural products. The average tariff on the imported agricultural products, as one of the trade barriers for the agricultural goods importing country, affects the virtual water flows. Furthermore, the irrigation water costs should be considered as a determinant. However, the irrigation prices are distorted by agricultural policies and are usually kept artificially low, which directly affects the farmers' decisions to produce (Wichelns 2004).

The productivity level of the agricultural sector influences the amount of virtual water flows. Some factors such as the number of tractors per 100 km^2 of arable land and kg of fertilizer consumed per hectare of arable land indicate the mechanization degree of a country and the level of agricultural productivity, as it is related to the agricultural sector (Fracasso *et al.* 2016). The education level of people in a country, based on the average number of years of education, reflects the level of socio-economic development and is considered as one of the factors influencing the virtual water flows (Barro & Lee 2013).

The model averaging method is used to select the determinants due to the high degree of uncertainty available in the empirical selection of the determinants of virtual water flows. The main features of the model averaging techniques are combining the model uncertainty with the analysis, averaging across all possible model characteristics, and identifying a set of possible explanatory variables by the researcher. Thus, the model averaging approach examines all possible sets of determinants instead of determining the correlation between the determinants based on one model or several pre-selected models. In the present study, the variables of population, GDP, average annual temperature, annual freshwater withdrawal for agriculture, arable land, and kg of fertilizer consumed per hectare were considered as the determinants of virtual water flows using model averaging. The gravity model was used to examine the factors affecting the virtual water flows of the selected crops and evaluate the export and import potential of these crops between Iran and other countries involved in importing or exporting these crops.

Evaluation of the gravity model

The empirical structure in this paper is based on a gravity model. The gravity model can accurately explain the bilateral trade flows. It is used for various purposes especially to assess market access, trade resistance, and impacts of regional and international agreements. The originality of the gravity model came back to Newton's law of gravitation in the 17th century. The gravity model is a function of attraction factors such as 'economic mass' (GDP) and resistance factors such as distance 'economic centers of gravity', population and trade preference factors like a common language, common borders, and others which can affect international trade (Ashktorab & Shahnoushi 2013).

The gravity model is widely used by economists as one of the most useful tools in interpreting and explaining the trade flows among the countries around the world. The gravity model is often estimated for the sum of exports and imports. Here, the data were used to evaluate the effects of various variables on exporting and importing the selected agricultural products based on the gravity model. The process is modeled as follows:

$$\mathrm{EX}_{it} = e^{\alpha_i} \prod\nolimits_m x_{imt}^{\beta m} \prod\nolimits_k e^{\gamma k P_{ikt}} \varepsilon_{it} \tag{1}$$

Since the structure of the gravity model is based on the linear logarithm, the coefficients of this model are in terms of elasticities or percent ratios. Such evaluations are comparable across countries and goods and directly reflect the trade responses to the potential trade variables.

In the form of linear logarithm,

$$\ln \mathrm{EX}_{it} = \alpha_i + \sum\nolimits_m \beta m \ln x_{imt} + \sum\nolimits_k {}^{\gamma k^{P_{ikt}}} + \ln \varepsilon_{it} \tag{2}$$

where EX_{it} indicates the amount of virtual water flows of selected Iranian agricultural products to/from country i in year t, x_{imt} demonstrates the mth explanatory variable for country i in year t, P_{ikt} represents the kth virtual variable, ε_{it} displays the error term, and α_i, β_m, and γ_k illustrate the parameters which are estimated.

The explanatory variables of the model consisted of the determinants of virtual water flows such as population of each country, GDP of each country, annual average temperature, annual freshwater withdrawal for agriculture (% of the total freshwater withdrawal), arable land per hectare, fertilizer consumption (kg per hectare of the arable land), the geographical distance between the capitals of Iran and ith country (km) reflecting the transportation cost and the amount of information of the two business partners, the virtual variable of the existence of a common border, and a water frontier between Iran and its trading partners. Number 1 indicates its existence. Otherwise, the number is 0.

The gravity model for the virtual water flows of the selected agricultural products of Iran is as follows:

$$\begin{aligned}\ln(\mathrm{EX}_{it}) = {} & \alpha_0 + \alpha_1 \ln(\mathrm{GDP}_{it}) + \alpha_2 \ln(\mathrm{POP}_{it}) + \alpha_3 \ln(\mathrm{AFW}_{it}) \\ & + \alpha_4 \ln(\mathrm{AL}_{it}) + \alpha_5 \ln(\mathrm{FCt}) + \alpha_5 \ln(\mathrm{Dis}_{ij}) + \alpha_6 \ln(\mathrm{TEM}_{it}) + \alpha_7 \mathrm{border} + \ln\varepsilon_{it}\end{aligned} \tag{3}$$

where EX_{it} displays the amount of virtual water flows of selected Iranian agricultural products to/from country i in year t, GDP_{it} represents the GDP of country i in year t, POP_{it} indicates the population of country i in year t, AFW_{it} demonstrates the annual freshwater withdrawal for agricultural purposes of country i in year t, AL_{it} illustrates the hectare of arable land in country i in year t, FC_t displays the kg per hectare of fertilizer consumed in Iran in year t, Dis_{ij} represents the distance between Iran and countries importing/exporting the selected agricultural products, TEM_{it} demonstrates the average annual temperature in country i in year t, border indicates the common border of Iran with the countries importing/exporting the selected agricultural products, and ε_{it} indicates the error term.

The selected agricultural products include wheat, barley, maize, rice, sugar beet, and tomato, as each is separately exported and imported from/to different countries during 2000–2016.

Predicting future virtual water flows under the climate and population change scenarios

Global warming due to human activities and climate changes is considered as one of the issues which have attracted the attention of many experts and scientists in different sciences. The global climate change caused by rising greenhouse gas concentrations, especially carbon dioxide, is expected to cause changes in precipitation regime, wind speed, sunlight reaching the surface, and air temperatures. Rising temperatures and climate change have detrimental effects on water resources. The climate change influences the water demand caused by rising temperatures and water supply (equilibrium in increasing carbon dioxide, evapotranspiration, and precipitation) (Hashempour *et al.* 2020). Therefore, adapting and coping with the climate change in the water resources sector along with their reduction reflection should be seriously considered as a part of the comprehensive response to the climate change-related vulnerability (Munang *et al.* 2013).

Currently, the most valid tool for generating climate change scenarios are the atmosphere-ocean coupled general circulation (AOGCM) 3D models (Wilby & Harris 2006), which use the periodic mean of AOGCM data in the climate change calculations. The early studies indicated that the best statistical period is 30 years. Based on the World Meteorological Organization (WMO), the base period of 1961–1990 is considered to coordinate and compare the base period selection in the different climate change studies. Furthermore, the WMO recommended to replace the period 1971–2000 when the data recorded at the stations of the studied area are not available for this period (IPCC-TGCIA 1999). Therefore, given the inventory of the climate variable statistics of the studied stations, the period of 1971–2000 was selected as the base period and the periods of 2016–2045 and 2070–2099 as the future periods.

The present study used the results of the study of Delghandi & Moazenzadeh (2017) as the output analysis of 15 models of HadCm3, CHAM5/MPI-OM, CSIRO-MK3.5, GFDL-CM2.0, CNRM-CM3, CGCM3 (T63), GISS-R, MRI-CGM2.3.2a, IPSL-CM4, CHO-G, BCM2.0 and MIROC3.2 mdrs, INMCM3.0, NCAR-CCSM3, and NCARPCM were used, which simulates the climate data under the A2, A1B, and B1 emission scenarios, indicating the most critical, average, and lowest rate of greenhouse gas emissions, respectively.

The results of their study indicated that the temperature changes in the first (2016–2045) and second future periods (2070–2099) compared to that in the previous period were calculated for Iran under the A2, A1B, and B1 emission scenarios using the existing equations. The middle number assigned to each period (2025 and 2085) was used to indicate the future period.

In the first future period, the three emission scenarios predicted the same temperature rise of about 1.5. However, it is different for the second future period when the pessimistic scenario of A2 and the optimistic scenario of B1 predicted the highest and lowest temperature increase, respectively. Furthermore, the A1B scenario is the intermediate of these two scenarios. In the second future period and under the A2 scenario, the temperature rises about 5 °C, since such temperature rises makes life difficult in the tropical regions of Iran, especially in the cities located in the margin of the Persian Gulf. The temperature rise is about 2.5 °C under the B1 scenario and about 4 °C under the A1B scenario (Delghandi & Moazenzadeh 2017).

In addition, the population of Iran as a determinant of the virtual water flows has been increasing over time. The average annual population growth is predicted by Statistical Center of Iran (2018). Based on the results, the average population growth is 0.98% for 2025 and 0.44% for 2085. Considering the average population growth of Iran in 2025 and 2085 as well as various emission scenarios, the net variations of the virtual water flows were calculated.

The data required for 2000–2016 such as population of each country, GDP of each country, annual freshwater withdrawal for agriculture (% of the total freshwater withdrawal), arable land (hectare), and fertilizer consumption (kg per hectare of arable land) were collected from the World Bank website. To provide the export/import volume of the selected products, destinations and the volume of export and import of each product were first identified from the website of the Food and Agriculture Organization of the United Nations (FAO). The water footprint of the selected products in Iran, calculated by Ashktorab (2019), was then used and multiplied by the amount of the export/import of the intended product to calculate the virtual water flows of each product. The average annual temperature of each country was obtained from the website of the National Centers for Environmental Information. The Internet software of city distance tool was used to determine the geographical distance between Iran and the capital of other countries. Accordingly, the geographical distance (km) of Tehran as the capital of Iran and the capital of the country importing/exporting the selected products was obtained, indicating the distance between these two countries.

Data

Population, GDP, annual freshwater withdrawal for agricultural purposes (% of the total freshwater withdrawal), arable land (hectare), and fertilizer consumption (kg per hectare of arable land) data for 2000–2016 were collected from the World Bank website. To provide the quantity of virtual water exports of six selected products, the destinations and the quantity of exports were identified from the Food and Agriculture Organization of the United Nations website, and the average annual temperature of each country was obtained from National Centers for Environmental Information. The average water footprint (Ashktorab 2019) of each product was used to calculate the virtual water flows.

RESULTS AND DISCUSSION

Given the equations, the results of tests, calculations, regression, and model prediction are presented, and based on the results, the final analysis and interpretation are performed.

Results of poolability test

The *F*-test was calculated separately for the cross-sections and time series. The obtained statistical value was 593.43. Based on the statistical value in the *F*-statistic table, H0 was rejected and the group effects were accepted. Therefore, the usability of the panel data method was confirmed.

The value of *F*-statistic for the period 2000–2016 was 742.12 and given the statistical value existing in the *F*-statistic table, H0 was rejected and applying the panel data method was confirmed.

Results of panel unit root test

The stationarity of the variables should first be examined in estimating the panel data models like the time-series models, and the effects of the random process are eliminated from the data if necessary. Given the already mentioned statistics, Table 2 indicates the panel unit root test results of the variables.

As shown in Table 2, all the variables reject the null hypothesis of the unit root in their levels by test: the Levin, Lin, and Chu tests. This finding indicates that all the variables are stationary.

Result of the Hausman test

The Hausman test statistic is equal to 10.56. Due to the calculated statistic, the H0, indicating the lack of correlation between individual effects and X_{it}, was rejected. Based on the results, the rejected random effects and the fixed effects are significant.

Results of the gravity model estimation with the panel data method

The net virtual water flows of the selected Iranian products were calculated based on each business partner during 2000–2016 (Figure 2).

The results indicated that Iran had the highest amount of virtual water imports for the selected products from Brazil, as well as the lowest amount of virtual water imports from Turkey during 2000–2016. Australia, South Africa, and Canada ranked the second to fourth as the highest exporters. The results revealed that most imports of the selected products and the virtual water were from long-distance countries and neighboring countries of Iran played a slight role in exporting these products to Iran during the 17-year period. Therefore, it is predicted that the distance variable, as a virtual water flow determinant, cannot have a significant effect on the virtual water flows of the selected products.

Table 2 | Results of panel unit root test

Variables	Levin–Lin & Chau test
$\ln(EX_{it})$	−5.88 (0.00)
$\ln(GDP_{others,t})$	−4.78 (0.00)
$\ln(GDP_{Iran,t})$	−7.52 (0.00)
$\ln(POP_{Iran,t})$	−5.59 (0.00)
$\ln(POP_{others,t})$	−4.36 (0.00)
$\ln(AL_{others,t})$	−13.88 (0.00)
$\ln(AL_{Iran,t})$	−1.58 (0.05)
$\ln(FCt)_{Iran}$	−2.96 (0.00)
$\ln(AFW_{others,t})$	−2.84 (0.00)
$\ln(AFW_{Iran,t})$	−1.74 (0.00)
$\ln(Dis_{ij})$	−7.01 (0.00)
$\ln(TEM_{others,t})$	−4.21 (0.00)
$\ln(TEM_{Iran,t})$	−3.36 (0.00)

Note: The value in parenthesis indicates the significance at 1% level with respect to the rejection of the null of nonstationary.

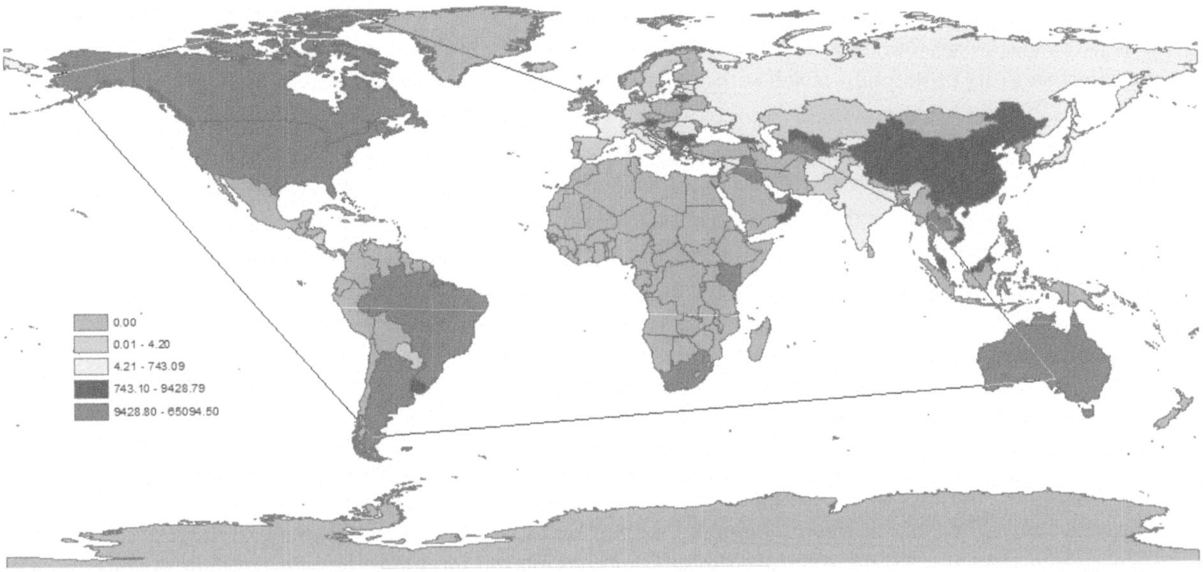

Figure 2 | Net virtual water flows of selected Iranian products with its business partners during 2000–2016 (million cubic meters).

Table 3 demonstrates the model estimation results to examine the effect of the determinants of the virtual water flows through the trade of the selected products using the panel data method.

The GDP of Iran and its trading partners was considered to estimate the gravity model of virtual water flows of selected products of Iran. An increase in the income and production of each country leads to an increase in its demand for imports of goods and services from the opposite country. Therefore, the coefficient of these variables is positive in the model. Table 3 indicates that the GDP of Iran is positive. In constant conditions, a 1% increase in the GDP of Iran enhances the virtual water flows of selected products by 0.29%. Since Iran is a net importer of the selected products (Ashktorab 2019), this coefficient demonstrates an increase of 29% in imports of these products. Based on the data obtained from the Food & Agriculture

Table 3 | Determinants of virtual water flows of selected agricultural products in Iran

Variables	Coefficients	t-statistic	Prob.
C	−4.38	−14.49	0.08
Population (Iran)	0.25	1.72	0.02
Population (partners)	−0.06	−3.03	0.09
GDP (Iran)	0.29	3.61	0.00
GDP (partners)	1.35	0.95	0.79
Arable land (Iran)	−0.36	−1.01	0.54
Arable land (partners)	0.12	3.49	0.00
Fertilizer consumption (Iran)	0.03	1.79	0.03
Annual freshwater withdrawal (Iran)	0.15	2.27	0.00
Annual freshwater withdrawal (partners)	−0.07	−1.61	0.01
Annual average temperature (partners)	−0.17	−1.26	0.82
Annual average temperature (Iran)	0.11	2.21	0.02
Distance	−0.79	−1.04	0.52
Common border	0.43	1.58	0.04
R^2	0.86	F-statistic	226.37
$\tilde{R}^2$	0.75	Durbin–Watson statistic	2.06

Organization (2016), Iran is a net importer of wheat, barley, maize, and rice and, consequently, a net importer of virtual water through these products. Given that these products have a significant share in the food pattern of Iran and wheat is a strategic product, the purpose of its production is self-sufficiency. Furthermore, the products such as rice and maize require a high amount of water. Considering the water resources status in Iran, the aim is to reduce their consumption. Therefore, the imports of these products enhance by increasing the GDP, and consequently, the virtual water imports increase through these products. On the other hand, it can be said that the economic growth caused by the increase in GDP leads the country to focus more on industrialization and shrink the agricultural sector (Figure 3). This factor causes a decrease in the export of agricultural products, which leads to a decrease in the export of virtual water, and in order to feed the population, the import of agricultural products, as well as the import of virtual water increase, while the time trend of this issue depends on the preferred policies of each country.

In this model, the population was separately determined for Iran and the countries importing/exporting selected products. It is predicted that the importing country with larger population imports more to meet their needs. However, the effect of the population increase in the exporting country on the imports of the importing country is unclear. Therefore, the coefficient of this variable seems positive for the population of the importing country, while this coefficient for the population of the exporting country is uncertain. As shown in Table 3, the population of Iran and the trading partners of the selected products were significant in the present study. The positive sign of the population of Iran indicates that the imports of these products increase when the population of Iran increases (Figure 4). Since the selected agricultural products are considered as the basic products in the food pattern of Iran, the population growth results in increasing the demand for these products, and accordingly the imports due to the reduction of the water resources. On the other hand, the exporting countries with a larger population are less inclined to the bilateral trade and focus more on domestic consumption. Furthermore, the results indicated that a 1% increase in the population leads to a 0.25% increase in the virtual water trade of the selected agricultural products in Iran. However, a 1% increase in the population of trading partners of Iran leads to a 0.06% reduction in the trade of the selected crops.

In the present study, the physical distance between countries is considered as a substitution for the transportation costs. Since the transportation costs have a negative impact on the trade between countries, the distance is negative based on the theory and has no significant effect on virtual water flows (Table 3). Given that the commercial targets of the selected crops are mainly in Europe and sometimes in Australia and Canada, the comparative advantage principle in relation to the crop production and the production technology leads to the reduction of the production costs and the final product prices, which covers the transportation costs. Therefore, Iran imports a greater share of its required selected crops from the countries with high physical distance (Figure 2). The common boundary is entered into the model as a virtual variable. In the present study, the countries with a common water or soil boundary with Iran are numbered 1 and the other countries as 0. Furthermore, the common border has a significant effect on the virtual water flows of the selected crops. Iran imports a

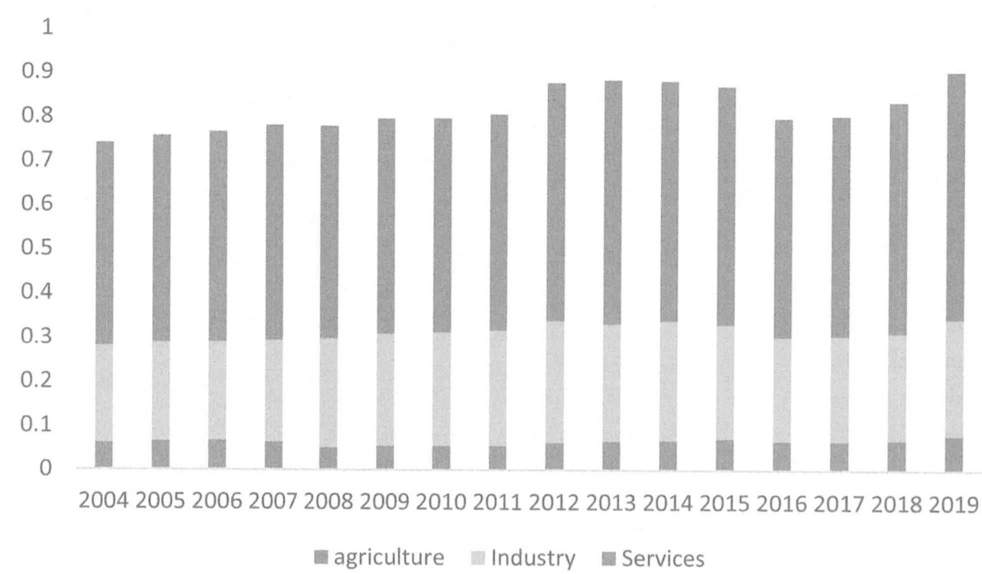

Figure 3 | Share of economic sectors in the GDP of Iran.

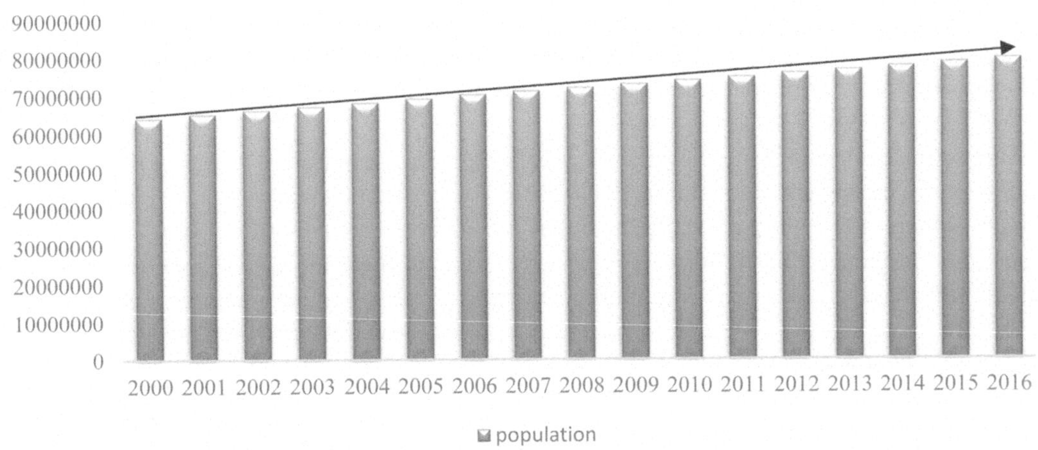

Figure 4 | Population of Iran during 2000–2016.

limited number of selected crops from neighboring countries and exports some of these crops to them. Arable land, fertilizer consumption, and annual freshwater withdrawal for agriculture affect virtual water flows. The arable land refers to the lands used for agriculture. This variable has a positive coefficient for exporting countries and a negative coefficient for importing countries. Table 3 demonstrates a negative coefficient for Iran, the net importer of selected agricultural products, which is not significant. Given that the water resources have declined dramatically in Iran in recent years, it is not possible to increase the production and reduce the imports of selected crops in Iran only by increasing the cultivation area. The low productivity of the arable land in Iran is related to the salinity, gravel, soil density, and lack of organic resources, impeding production. The coefficient of an arable land variable is positive and significant for trading partners of Iran, which indicates that the trading partners of Iran increase the production of the desired crops by increasing cultivated lands, leading to an increase in the exports of these crops and virtual water. There are several variables indicating the pressure on the water resources (Fracasso *et al.* 2016). In the same vein, the annual freshwater withdrawal variable for agriculture is considered as one of the variables demonstrating the pressure on the renewable water resources. The virtual water trade decreases in the countries with high water resources withdrawal. Therefore, this variable has a negative coefficient for the countries exporting the virtual water and a positive coefficient for the virtual water importing countries.

As shown in Table 3, the annual freshwater withdrawal variable for agriculture in Iran is positive and significant, indicating that the pressure on the renewable water resources in Iran decreases by increasing the virtual water trade. This coefficient is negative and significant for the trading partners in Iran, which demonstrates that the export of agricultural products increases the pressure on the water resources of those countries. A 1% increase in the annual freshwater withdrawal for agriculture enhances the virtual water imports of Iran by 0.15% due to the increased water withdrawal costs and reduces the virtual water exports of trading partners by 0.07%, since an increase in groundwater utilization leads to irreparable damages to the groundwater resources. The excessive withdrawal of groundwater resources increases the saline aquifers, as well as the salinity in the freshwater resources in different regions of the world. The level of underground aquifers has decreased in Iran over the past decades due to the continuation of low precipitation compared to the long-term trend and the improper utilization of water resources, especially in the agriculture (Karimi *et al.* 2018). In this regard, the virtual water imports reduce the pressures on the water resources.

Fracasso *et al.* (2016) considered the fertilizer consumed per hectare of arable land as a measure of the mechanization of a country. This variable is intended only for the country of origin, which positively affects the virtual water flows, which has a positive relationship with the production technology. Based on the results in Table 3, the coefficient of this variable is positive and significant, demonstrating that the enhancement of the fertilizer consumed per hectare results in increasing the production, as well as the exports of agricultural products and virtual water. Although the use of agricultural fertilizers increases the gray water footprint, their impact on increasing production cannot be ignored. Furthermore, the changes in the fertilizer consumed per hectare in Iran alter the virtual water flows of the selected crops by 3%. Based on the average annual temperature, the temperature and heat enhancement increase the evapotranspiration and the water requirement of the plant and, subsequently, alter the water footprint of the plant and the virtual water flows. The positive coefficient of this variable for Iran demonstrates that the net virtual water flows enhance by increasing the average annual temperature. Furthermore, a 1% increase in the average annual temperature enhances the virtual water imports of Iran by 0.11%, due to the increased evapotranspiration at the farm level. The coefficient of this variable is insignificant for other countries (Table 3).

Figure 5 demonstrates the net virtual water flows of the selected products separately. The net virtual water flows of wheat, barley, maize, and rice have the highest value during 2000–2016 (Figure 5). The average population growth in Iran in the 10-year period before 2014 increased more than that in the other periods (Statistical Center of Iran 2018). Therefore, the population growth leads to a net increase in the imports of selected crops to meet the demands of the growing population. In addition to population changes, an increase in rainfall in some years has led to an increase in green water footprint, and also in some years, a decrease in rainfall has led to a decrease in green water footprint. Changes in water footprint have led to fluctuations in net virtual water flows.

Future virtual water flows under different scenarios

Future virtual water flows under climate change scenarios

The present study predicted the size of the virtual water flows in the future based on the findings of Delghandi & Moazenzadeh (2017), in relation to the average annual temperature change in the coming years under the A2, A1B, and B1 emission scenarios, indicating the most critical, average, and lowest greenhouse gas emissions, respectively.

The determinants of virtual water flows were identified during 2000–2016. The average annual temperature as a determinant of virtual water flows has a positive and significant effect on the net virtual water flows of selected crops in Iran (Table 3). The other variables were considered constant, and the changes were applied to the average annual temperature in Iran to predict the changes in the virtual water flows in the future (Table 4).

In the first future period (2016–2045), the average annual temperature under three climate change scenarios increases by 1.5 °C, leading to a change of 6.1 milliard cubic meters in net virtual water flows. Given that an average of 37 milliard cubic meters of virtual water was imported through importing the selected products during 2000–2016 (Ashktorab 2019), this

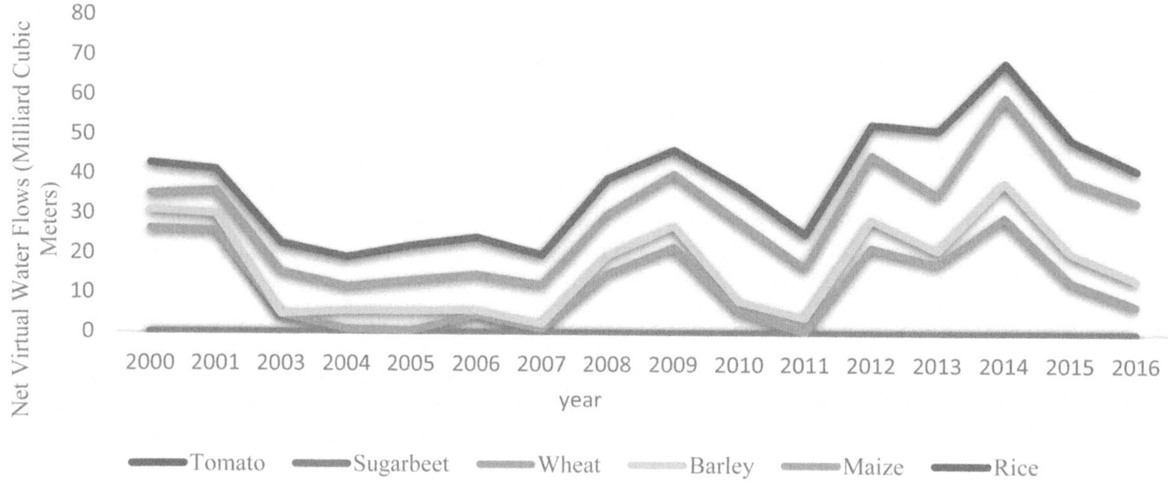

Figure 5 | Net virtual water flows of selected products (2000–2016).

Table 4 | Predicting virtual water flow changes under climate change scenarios

	Climate change scenarios		
	A2	**B1**	**A1B**
Net virtual water flow changes in the first future period (milliard cubic meters)	6.1	6.1	6.1
Net virtual water flow changes in the second future period (milliard cubic meters)	20.33	10.16	16.26

amount increased by 6.1 milliard cubic meters in the first future period due to the positive effect of the change in the average annual temperature on the virtual water flows (Table 4).

In the second future period (2070–2099), the average annual temperature under scenario A2 (pessimistic) increases by 5 °C, leading to an increase of 20.33 milliard cubic meters of the virtual water imports through importing the selected products. In the scenario B1 (optimistic), a 2.5 °C increase in the average annual temperature enhances 10.16 milliard cubic meters of virtual water flows. Furthermore, a 4 °C increase in the temperature in the scenario A1B results in increasing 16.26 milliard cubic meters in the net virtual water flows in the second future period (Table 4).

Future virtual water flows under the population change scenarios

In this section, the virtual water flows in the future under the population change scenarios were predicted. The average annual population growth of Iran for 2025 and 2085 is calculated 0.98 and 0.44%, respectively.

The average population of Iran, as a determinant of virtual water flows, has a positive and significant effect on the net virtual water flows of the selected crops in Iran (Table 3). To predict changes in the future virtual water flows, other variables were considered constant, and the changes were applied to the average population of Iran (Table 5).

In the first future period (2016–2045), the average population growth is 0.98%, which leads to a change of 9.1 milliard cubic meters in net virtual water flows. Considering that an average of 37 milliard cubic meters of the virtual water was imported through importing the selected products during 2000–2016, this amount increased to 9.1 milliard cubic meters in the first period due to the positive effect of changing the average annual temperature and the population of Iran on the virtual water flows (Table 5). In the second future period (2070–2099), the average population growth is 0.44%, leading to an increase of 4.07 milliard cubic meters of virtual water imports through importing the selected products (Table 5).

Future virtual water flows under the climate and population change scenarios

The virtual water flows in the future under the climate and population change scenarios were predicted here. The average annual temperature and the population of Iran, as the determinants of virtual water flows, have a positive and significant effect on the net virtual water flows of the selected crops in Iran (Table 3). The average annual temperature and the population of Iran were changed, and other variables were considered constant to predict changes in the future virtual water flows (Table 6).

In the first future period (2016–2045), the average annual temperature under all three climate change scenarios increases by 1.5 °C and the average population growth is 0.98%, leading to a change of 15.17 milliard cubic meters in the net virtual water flow (Table 6).

In the second period (2070–2099), the average annual temperature under scenario A2 (pessimistic) increases by 5 °C and the average population growth is 0.44%, leading to an increase of 24.42 milliard cubic meters of virtual water imports by

Table 5 | Predicting virtual water flow changes under population change scenarios

	Population change scenarios
Net virtual water flow changes in the first future period (milliard cubic meters)	9.1
Net virtual water flow changes in the second future period (milliard cubic meters)	4.07

Table 6 | Predicting virtual water flow changes under climate and population change scenarios

		Climate change scenarios		
	Population change scenarios	A2	B1	A1B
Net virtual water flow changes in the first future period (milliard cubic meters)	The average population growth is 0.98%	15.17	15.17	15.17
Net virtual water flow changes in the second future period (milliard cubic meters)	The average population growth is 0.44%	24.42	14.25	20.35

importing the selected products. In the scenario B1 (optimistic), an increase of 2.5 °C in the average annual temperature and the average population growth of 0.44% leads to an increase of 14.25 milliard cubic meters of the net virtual water flows. Furthermore, in the A1B scenario, an increase of 4 °C, along with an average population growth of 0.44%, increases the temperature to 20.35 milliard cubic meters of the net virtual water flows in the second future period (Table 6).

Based on the results, the temperature and population variation in the future change the amount of the virtual water flows. Therefore, the imports of agricultural products increase in future periods to meet the food demand of the population, leading to the increased virtual water imports.

CONCLUSION

Given that agriculture is considered as the largest consumer of water resources in the world, the management of the virtual water trade is proposed as a policy for food supply, protection, and sustainability of water resources, as well as a strategy for the storage of water resources of each country, and achieving water security at the national level. In the present study, the determinants of virtual water flows were identified and their influence was examined using the gravity model. Furthermore, the size of the future virtual water flows was predicted under the climate and population change scenarios. The results indicated that the factors such as the populations of countries having trade flows (exports and imports) together, GDP, annual freshwater withdrawal for agriculture (% of the total freshwater withdrawal), arable land in hectare, fertilizer consumption in kg per hectares of arable land, the distance between countries, and the common border are the determinants of virtual water flows in Iran. Based on the results of the Food and Agriculture Organization (2016), Iran is a net importer of wheat, barley, maize, and rice and, consequently, a net importer of virtual water through these products. Given that the water resources in Iran have declined dramatically in recent years, the production cannot be increased and the imports of the selected agricultural products cannot be reduced in Iran only by increasing the cultivation area.

The early domestic studies, which examined temperature changes under the climate change scenarios, were used to predict the size of the virtual water flows of Iran in the future. The results indicated that the average annual temperature under all three climate change scenarios increase by 1.5 °C in the first future period (2016–2045), which can lead to a change of 6.1 milliard cubic meters in the net virtual water flow. In the second future period (2070–2099), the average annual temperature under the pessimistic, optimistic, and intermediate scenarios can increase by 5, 2.5, and 4 °C, respectively, leading to an increase of 20.33, 10.16, and 16.26 milliard cubic meters of virtual water imports through the import of the selected products. Furthermore, the climate change scenarios, along with the average population change, were examined in both future periods. The average population growth in the first and second future periods was 0.98 and 0.44%, respectively. The results indicated that the net virtual water flows increase by 9.44 milliard cubic meters in the first future period. In addition, the net virtual water flows increase by 21.82, 11.64, and 17.75 milliard cubic meters, respectively, in the second future period under the three pessimistic, optimistic, and intermediate scenarios.

By considering the above-mentioned issues, as well as the political conditions prevailing in the country, the following recommendations are suggested:

1. Increasing the virtual water flows through the imports of the agricultural products with high water footprint results in increasing the share of flows to the irrigation water in the agricultural sector and decreasing the pressure on the water resources of Iran. Therefore, it is recommended to change the trade policies fundamentally based on the enhancement of the virtual water imports.
2. The results indicated that the agricultural products are mainly imported from countries with a long geographical distance from Iran. Therefore, it is suggested to decrease the import tariff of agricultural products with high water footprint and increase the export tariff of agricultural products with high water footprint to increase the virtual water imports. On the other hand, finding similar products from nearby or neighboring countries that have a lower water footprint is also a real possibility in virtual water trading.
3. The government can use an *ad valorem* import and export tariff policy depending on the amount of each product's water footprint so that products with more water footprint apply higher export tariffs as well as lower import tariffs.
4. The government can use export quotas according to the water footprint, which could lead to a reduction in the area under cultivation of water-intensive crops.

5. The results of the simulation scenario of population and temperature enhancement demonstrated an increase in the virtual water imports into Iran. Therefore, it is necessary to identify countries with the lowest import costs and adopt appropriate trade policies as soon as possible.
6. The food supply for future generations is considered as one of the problems facing policymakers due to the water resources constraints. Given the political relations of Iran with other countries, the population reduction policies should be considered over a specific period.
7. Considering that the dominant food pattern in Iran includes rice as a water-consuming product, changing the present food pattern significantly reduces the pressure on the water resources, as well as the dependence on imports of this product.

FUNDING

This research did not receive any specific grant from funding agencies in the public, commercial, or not-for-profit sectors.

DATA AVAILABILITY STATEMENT

All relevant data are included in the paper or its Supplementary Information.

REFERENCES

Allan, J. A. 1993 *Fortunately, There Are Substitutes for Water Otherwise Our Hydropolitical Futures Would Be Impossible.* Priorities for Water Resources Allocation and Management. ODA, London, UK.

Allan, J. A. 1998 Virtual water: a strategic resource global solution to regional deficits. *Ground Water* **36**, 545–546.

Aquastat 2017 *FAO Aquastat Country Profile.* Available from: http://www.fao.org/aquastat/en/countries-and-basins/country-profiles/.

Ashktorab, N. 2019 *Virtual Water Flows and Determinants in Iran.* PhD Thesis. Shiraz University, Shiraz, Iran.

Ashktorab, N. & Shahnoushi, N. 2013 Evaluating the effect of aflatoxin food safety concerns on Iran's pistachio trade with gravity model. *Afr. J. Agric. Res.* **8** (43), 5320–5326. doi:10.5897/AJAR11.1952.

Barro, R. & Lee, J. W. 2013 A new data set of educational attainment in the world, 1950–2010. *J. Dev. Econ.* **104**, 184–198. https://doi.org/10.1016/j.jdeveco.2012.10.001.

Biewald, A., Rolinski, S., Camoen, H., Schmitz, C. & Dietrich, J. 2014 Valuing the impact of trade on local blue water. *Ecol. Econ.* **101**, 43–53. https://doi.org/10.1016/j.ecolecon.2014.02.003.

Casolani, N., Pattara, C. & Liberator, L. 2016 Water and carbon footprint perspective in Italian durum wheat production. *Land Use Policy* **58**, 394–402. https://doi.org/10.1016/j.landusepol.2016.07.014.

Chapagain, A. K., Hoekstra, A. & Savenije, H. 2006 Water saving through international trade of agricultural products. *Hydrol. Earth Syst. Sci.* **10**, 455–468. https://doi.org/10.5194/hess-10-455-2006.

Chouchane, H., Krol, M. S. & Hoekstra, A. Y. 2018 Virtual water trade patterns in relation to environmental and socioeconomic factors: a case study for Tunisia. *Sci. Total Environ.* **613–614**, 287–297.

Dalin, C. 2014 *Water for Food: Evolution and Projections of Water Transfers Through International and Domestic Agricultural Trade.* PhD Thesis. Princeton University, Princeton, NJ, USA.

DaMatta, F. M., Grandis, A., Arenque, B. C. & Buckeridge, M. S. 2010 Impacts of climate change on crop physiology and food quality. *Food Res. Int.* **43**, 1814–1823. https://doi.org/10.1016/j.foodres.2009.11.001.

Debaere, P. 2014 The global economics of water: is water a source of comparative advantage? *Am. Econ. J. Appl. Econ.* **6** (2), 32–48. https://doi.org/10.1257/app.6.2.32.

De Fraiture, C., Cai, X., Amarasinghe, U., Rosegrant, M. & Molden, D. 2004 *Does International Cereal Trade Save Water? The Impact of Virtual Water Trade on Global Water Use.* International Water Management Institute (IWMI), Comprehensive Assessment of Water Management in Agriculture Research Report 004, Colombo, Sri Lanka. http://dx.doi.org/10.3910/2009.389.

Delghandi, M. & Moazenzadeh, R. 2017 Investigating spatiotemporal variations of precipitation and temperature over Iran under climate change condition considering AOGCM models and emission scenarios uncertainty. *Iran. J. Ecohydrol.* **3** (3), 321–331.

Deng, K., Yu, L., Zheng, X., Zhang, K., Wang, W., Dong, P., Zhang, J. & Ren, M. 2016 Target of rapamycin is a key player for auxin signaling transduction in arabidopsis. *Front. Plant Sci.* **7**, 291. doi:10.3389/fpls.2016.00291.

Ercin, A. E. & Hoekstra, A. Y. 2014 Water footprint scenarios for 2050: a global analysis. *Environ. Int.* **64**, 71–82. https://doi.org/10.1016/j.envint.2013.11.019.

Fader, M., Rost, S., Muller, C., Bondeau, A. & Gerten, D. 2010 Virtual water content of temperate cereals and maize: present and potential future patterns. *J. Hydrol.* **384** (3–4), 218–231. https://doi.org/10.1016/j.jhydrol.2009.12.011.

Food and Agriculture Organization 2016 *Faostatical Year Book 2016, Data, Trade, Detailed Trade Matrix, Selected Crops 2016.*

Foresight 2011 *The Future of Food and Farming: Challenges and Choices for Global Sustainability.* Final Project Report. The Government Office for Science, London, UK.

Fracasso, A. 2014 A gravity model of virtual water trade. *Ecol. Econ.* **108**, 215–228. https://doi.org/10.1016/j.ecolecon.2014.10.010.

Fracasso, A., Sartori, M. & Schiavo, S. 2016 Determinants of virtual water flows in the Mediterranean. *Sci. Total Environ.* **543**, 1054–1062. https://doi.org/10.1016/j.scitotenv.2015.02.059.

Gleik, P. H. 2008 Can California's water problems be solved. *Ecol. Law Currents* **35** (71), 71–79.

Haddeland, I., Heinke, J., Biemans, H., Eisner, S., Flörke, M., Hanasaki, N., Konzmann, M., Ludwig, F., Masaki, Y. & Schewe, J. 2014 Global water resources affected by human interventions and climate change. *Proceedings of National Academy Sciences of the United States of America* **111** (9), 3251–3256.

Hanasaki, N., Fujimori, S., Yamamoto, T., Yoshikawa, S., Masaki, Y., Hijioka, Y., Kainuma, M., Kanamori, Y., Masui, T. & Takahashi, K. 2013 A global water scarcity assessment under shared socio-economic pathways – Part 2: water availability and scarcity. *Hydrol. Earth Syst. Sci.* **17** (7), 2393–2413. https://doi.org/10.5194/hess-17-2393-2013.

Hashempour, Y., Nasseri, M., Mohseni-Bandpei, A., Motesaddi, S. & Eslamizadeh, M. 2020 Assessing vulnerability to climate change for total organic carbon in a system of drinking water supply. *Sustain. Cities Soc.* **53**, 101904. https://doi.org/10.1016/j.scs.2019.101904.

Hoekstra, A. Y. & Hung, P. Q. 2002 *Virtual Water Trade: A Quantification of Virtual Water Flows Between Nations in Relation to International Crop Trade.* Value of Water Research Report Series No. 11.

Huang, S. L., Yeh, C. T. & Chang, L. F. 2010 The transition to an urbanizing world and the demand for natural resources. *Curr. Opin. Environ. Sustain.* **2** (3), 136–143. https://doi.org/10.1016/j.cosust.2010.06.004.

IPCC-TGCIA 1999 *Guidelines on the Use of Scenario Data for Climate Impact and Adaptation Assessment.* T. R. Carter, M. Hulme & M. Lal, eds. Intergovernmental Panel on Climate Change, Task Group on Scenarios for Climate Impact Assessment, pp. 1–69.

Islam, M., Oki, T., Kanae, S., Hanasaki, N., Agata, Y. & Yoshimura, K. 2007 A grid-based assessment of global water scarcity including virtual water trading. *Water Resour. Manage.* **21**, 19–33. doi:10.1007/s11269-006-9038-y.

Karimi, V., Karami, E. & Keshavarz, M. 2018 Climate change and agriculture: impacts and adaptive responses in Iran. *J. Integr. Agric.* **17** (1), 1–15. https://doi.org/10.1016/S2095-3119(17)61794-5.

Konar, M., Hussein, Z., Hanasaki, N., Mauzerall, D. L. & Rodriguez-Iturbe, I. 2013 Virtual water trade flows and savings under climate change. *Hydrol. Earth Syst. Sci.* **17** (8), 3219–3234. https://doi.org/10.5194/hess-17-3219-2013.

Kumar, M. & Singh, O. 2005 Virtual water in global food and water policy making: is there a need for rethinking? *Water Resour. Manag.* **19**, 759–789. https://doi.org/10.1007/s11269-005-3278-0.

Lenzen, M., Bhaduri, A., Moran, D., Kanemoto, K., Bekchanov, M., Geschke, A. & Foran, B. 2012 *The Role of Scarcity in Global Virtual Water Flows.* Discussion Papers 133478. University of Bonn, Center for Development Research (ZEF). https://doi.org/10.22004/ag.econ.133478.

Lenzen, M., Moran, D., Bhaduri, A., Kanemoto, K., Bekchanov, M., Geschke, A. & Foran, B. 2013 International trade of scarce water. *Ecol. Econ.* **94**, 78–85. https://doi.org/10.1016/j.ecolecon.2013.06.018.

Liu, J. G., Folberth, C., Yang, H., Rockstrom, J., Abbaspour, K. & Zehnder, A. J. B. 2013 A global and spatially explicit assessment of climate change impacts on crop production and consumptive water use. *PLoS One* **8** (2), e57750. https://doi.org/10.1371/journal.pone.0057750.

Mubako, S. T. 2011 *Frameworks for Estimating Virtual Water Flows among U.S. States.* PhD Thesis. University of Carbondale, Carbondale, USA.

Munang, R., Thiaw, I., Alverson, K., Muba, M., Liu, J. & Rivington, M. 2013 Climate change and ecosystem-based adaptation: a new pragmatic approach to buffering climate change impacts. *Curr. Opin. Env. Sust.* **5** (1), 67–71. https://doi.org/10.1016/j.cosust.2012.12.001.

Novo, P., Garrido, A. & Varela-Ortega, C. 2009 Are virtual water 'flows' in Spanish grain trade consistent with relative water scarcity? *Ecol. Econ.* **68** (5), 1454–1464. https://doi.org/10.1016/j.ecolecon.2008.10.013.

Oki, T. & Kanae, S. 2004 Virtual water trade and world water resources. *Water Sci. Technol.* **49** (7), 203–209. https://doi.org/10.2166/wst.2004.0456.

Oki, T., Yano, S. & Hanasaki, N. 2017 Economic aspects of virtual water trade. *Environ. Res. Lett.* **12**, 044002. https://doi.org/10.1088/1748-9326/aa625f.

Pfister, S., Bayer, P., Koehler, A. & Hellweg, S. 2011 Projected water consumption in future global agriculture: scenarios and related impacts. *Sci. Total Environ.* **409** (20), 4206–4216. https://doi.org/10.1016/j.scitotenv.2011.07.019.

Ramirez-Vallejo, J. & Rogers, P. 2004 Virtual water flows and trade liberalization. *Water Sci. Technol.* **49** (7), 25–32. https://doi.org/10.2166/wst.2004.0407.

Roson, R. & Sartori, M. 2010 *Water Scarcity and Virtual Water Trade in the Mediterranean.* University Ca' Foscari of Venice, Department of Economics Research Paper Series No. 08_10. http://dx.doi.org/10.2139/ssrn.1594709.

Roson, R. & Sartori, M. 2013 Analysis of the virtual water trade in Italy and in the mediterranean. In: *The Water We Eat. What Is Virtual Water and How We Use It* (Antonelli, M. & Greco, F., eds). Edizioni Ambiente, Frosinone, pp. 181–198.

Rost, S., Gerten, D., Bondeau, A., Lucht, W., Rohwer, J. & Schapho, S. 2008 Agricultural green and blue water consumption and its influence on the global water system. *Water Resour. Res.* **44** (9), 1–17. https://doi.org/10.1029/2007WR006331.

Statistical Center of Iran 2018 General Results of the Population and Housing Census of the Country, 2016, Iran, Tehran.

Strzepek, K. & Boehlert, B. 2010 Competition for water for the food system. *Philos. Trans. R. Soc. Lond. B* **365** (1554), 2927–2940. doi:10.1098/rstb.2010.0152.

Thomson, A., Hallsworth, S. & Malcolm, H. 2013 *Projections of Emissions and Removals From the UK LULUCF Sector to 2050. Contract Report Prepared for the Department of Energy and Climate Change (DECC) as Part of the Contract Inventory and Projections of UK*

Emissions by Sources and Removals by Sinks due to Land Use, Land-Use Change and Forestry (LULUCF). Centre for Ecology and Hydrology, 28.

United Nations 2015 *The United Nations World Water Development Report.* Water for a Sustainable World.

United Nations 2019 *World Population Prospects 2019.* Department of Economic and Social Affairs, Population Dynamics.

Verhofstadt, L., Devoldre, I., Buysse, A., Stevens, M., Hinnekens, C., Ickes, W. & Davis, M. 2016 The role of cognitive and affective empathy in spouses' support interactions: an observational study. *PLoS One* **11** (2), e0149944. https://doi.org/10.1371/journal.pone.0149944.

Verma, S., Kampman, D. A., Van der Zaag, P. & Hoekstra, A. Y. 2009 Going against the flow: a critical analysis of inter-state virtual water trade in the context of India's national river linking program. *Phys. Chem. Earth* **34** (4–5), 261–269. https://doi.org/10.1016/j.pce.2008.05.002.

Vorosmarty, C. J., Green, P., Salisbury, J. & Lammers, R. B. 2000 Global water resources: vulnerability from climate change and population growth. *Science* **289** (5477), 284–288. doi:10.1126/science.289.5477.284.

Wada, Y. & Bierkens, M. F. P. 2014 Sustainability of global water use: past reconstruction and future projections. *Environ. Res. Lett.* **9** (10), 104003. doi:10.1088/1748-9326/9/10/104003.

Wang, Z., Kai, H., Shunshun, Y. & Yajuan, Y. 2013 An input-output approach to evaluate the water footprint and virtual water trade of Beijing, China. *J. Clean. Prod.* **42**, 172–179.

Wichelns, D. 2004 The policy relevance of virtual water can be enhanced by considering comparative advantages. *Agric. Water Manag.* **66**, 49–63. https://doi.org/10.1016/j.agwat.2003.09.006.

Wilby, R. L. & Harris, I. 2006 A framework for assessing uncertainties in climate change impacts: low flow scenarios for the River Thames, UK. *Water Resour. Res.* **42** (2), W02419. https://doi.org/10.1029/2005WR004065.

World Bank 2014 *Annual Freshwater Withdrawals, Total (Billion Cubic Meters)*. Available from: http://data.worldbank.org/indicator/ER.H2O.FWTL.K3.

World Bank 2016 *Development Prospect Group (DEPG), Data and Statistics.*

Yang, H. & Zehnder, A. 2007 Virtual water: an unfolding concept in integrated water resources management. *Water Resour. Res.* **43** (12), 1–10. https://doi.org/10.1029/2007WR006048.

Yang, H., Reichert, P., Abbaspour, K. & Zehnder, A. 2003 A water resources threshold and its implications for food security. *Environ. Sci. Technol.* **37** (14), 3048–3054.

Yawson, D. O., Mohan, S., Armah, F. A., Ball, T., Mulholand, B., Adu, M. O. & White, P. J. 2020 Virtual water flows under projected climate, land use and population change: the case of UK feed barley and meat. *Heliyon*, e03127. https://doi.org/10.1016/j.heliyon.2019.e03127.

Zhang, Y., Zhang, J.-H., Tian, Q., Liu, Z.-H. & Zhang, H.-L. 2018 Virtual water trade of agricultural products: a new perspective to explore the belt and road. *Sci. Total Environ.* **622–623**, 988–996. https://doi.org/10.1016/j.scitotenv.2017.11.351.

Zhu, X. F., Zhao, A. Z., Li, Y. Z. & Liu, X. F. 2015 Agricultural irrigation requirements under future climate scenarios in China. *J. Arid Land* **7** (2), 224–237. doi:10.1007/s40333-014-0080-y.

Zhuo, L., Mekonnen, M. M., Hoekstra, A. Y. & Wada, Y. 2016 Inter and intra-annual variation of water footprint of crops and blue water scarcity in the Yellow River Basin (1961–2009). *Adv. Water Resour.* **87**, 21–41. https://doi.org/10.1016/j.advwatres.2015.11.002.

Zimmer, D. & Renault, D. 2003 Virtual water in food production and global trade: review of methodological issues and preliminary results. In: *Virtual Water Trade, Proceedings of the International Expert Meeting on Virtual Water Trade.* Delft, The Netherlands.

Zoumides, C., Zachariadis, T., Bruggeman, A. & Markou, M. 2014 *Policy Options and Economic Scenarios.* AGWATER Scientific Report 6. Nicosia, Cyprus.

First received 14 May 2021; accepted in revised form 16 August 2021. Available online 30 August 2021

Water Management Addressing Societal and Climate Change Challenges doi: 10.2166/wcc.2021.412

Exploring the village tank cascade systems (VTCSs) in Vavuniya district, Sri Lanka

A. Nanthakumaran [a,*], H. K. Kadupitiya [b], S. Devaisy[a] and W. E. P. Athukorale[a]

[a] Department of Bio-science, Faculty of Applied Science, Vavuniya Campus of the University of Jaffna, Vavuniya, Sri Lanka
[b] Department of Agriculture, Natural Resource Management Centre, Peradeniya, Sri Lanka
*Corresponding author. E-mail: n_ananthi@vau.jfn.ac.lk

AN, 0000-0003-0224-3957; HKK, 0000-0002-9300-0485

ABSTRACT

An attempt was made to identify, validate the village tank cascade systems (VTCSs) and study the water flow from one village tank to another in each VTCS in the eight Agrarian Service Centre (ASC) divisions in the Vavuniya district from October 2017 to December 2018. VTCS contribute a significant share of available water resources for the livelihoods of households in the Vavuniya district. The 1:10,000 topographic map of the Survey Department, satellite images and the digital elevation model were used to identify the cascades and flow direction map for the study area using ArcGIS 10.2.2. Among 756 village tanks in the district, 80 VTCSs comprising 514 village tanks were identified, and only 69 cascades were validated in the field. In addition, this study identified 111 isolated village tanks without connecting with other village tanks and 131 abandoned village tanks. Further investigation is recommended to explore the possibilities of increasing the cascade areas in the study area by connecting isolated tanks with VTCSs. Initiation taken towards rehabilitation of cascades would enhance the livelihood of farm households in the Vavuniya district and lead to sustainable water resource management.

Key words: cascades, DEM, GIS, livelihood, validation, village tanks

HIGHLIGHTS

- The VTCS play a crucial role in sustainable management and water conservation.
- The number of the identified VTCSs is more than that of validated in the field of the study area.
- Tank siltation and water scarcity altered the natural settings of the VTCS.
- The findings will benefit urban planners for developmental activities and farmers for their livelihood.

INTRODUCTION

The village tank cascade system (VTCS) is 'a connected series of village tanks organized within a micro-catchment of the dry zone landscape, storing, conveying and utilizing water from an ephemeral rivulet' (Madduma Bandara 1985). These systems are one of the oldest and the most advanced water conveyance/irrigation systems during the hydraulic civilization (Mahatantila *et al.* 2008; Nanthakumaran & Palanisami 2010; Ratnayake *et al.* 2021). In Sri Lanka, these systems were developed by the ancient kings during the Rajarata hydraulic civilization as a water conservation strategy by harvesting rainwater to optimize the water utilization for irrigation in the dry zone of Sri Lanka (Madduma Bandara 1985). Rainwater is the sole input to the VTCS; thus, these are considered rainwater harvesting structures. The excess water received by the upstream tank is allowed to spill out through sluices to flood the rice fields/other vegetation downstream, then the excess water from the rice fields/vegetation is captured by the adjacent downstream reservoir. Thus, the water resource is continuously recycled (Panabokke *et al.* 2002; Lakshman *et al.* 2019; Dharmasena 2020). Hence, the cascade lines are considered an adaptation strategy to overcome threats from the long-term changes due to climate change (Withanachchi *et al.* 2014; Galagedara *et al.* 2018).

VTCSs and the unconfined aquifer of groundwater are the available water resources for the livelihoods of households in the Vavuniya district. The Vavuniya district falls under the dry zone of Sri Lanka and receives less rainfall (annual average of <1,200 mm), high evaporation (mean annual of 1,700–1,900 mm) and high temperature (mean annual of 26 – 34 °C), resulting in hot weather and water shortage especially from May to August. Madduma Bandara (1985) reported that these VTCSs contributed to the shallow regolith aquifer recharge maintaining a high water table. Previous studies found substantial groundwater recharge in the proximity to VTCS operational in the Vavuniya district. Tharani *et al.* (2015) found a significant positive correlation between the water levels of the village tanks and the water levels of the wells situated in its command area. Shanmuganathan *et al.* (2010) identified that human alterations on the proximity to the cascades had reduced the sustainable production of groundwater resources in the localities.

Therefore, it is vital to keep the cascades operational to ensure the sustainable utilization of water resources. The VTCS must be considered in the planning stages of any development projects to ensure they are environmentally sound and economically viable in nature. In view of this, lack of proper documentation of the VTCS functioning in Vavuniya (Sudusinghe *et al.* 2016) is the major constraint for decision-makers to integrate ecological aspects in the environmental impact assessments.

Furthermore, it has been explored that the village tank performance had been declining with a reduction in tank capacity due to tank siltation and improper maintenance. This situation made the farmers request tank rehabilitation to store more water in the village tanks. In the recent past, the non-governmental organizations and the World Bank funded and rehabilitated some randomly selected village tanks in the Vavuniya district, as there is no documentation or knowledge of the cascade lines of the village tanks. This resulted in the inundation of adjacent tanks due to overflow of tanks during the rainy season (Personal Communication). It ultimately affected the livelihood of the farmers in the adjacent tanks. The objective of the study was to identify and validate the VTCS and to study the water flow from one village tank to another in each VTCS located in the eight ASC divisions in the Vavuniya district.

STUDY AREA

The whole Vavuniya district was considered for this study. The cascades were analyzed in the eight ASCs, namely *Cheddikulam*, *Pampaimadu*, *Kovilkulam*, *Madukandha*, *Nedunkerny*, *Omanthai*, *Ulukulam* and *Kanagarayankulam*, which are shown in Figure 1. The total extent of the study area is 1,995 km^2.

METHODS

The identification and mapping of VTCSs separately for each ASC in the Vavuniya district were performed through GIS approaches with available digital maps, remote sensing images, field investigation and validation in consultation with local experienced senior people in the respective village tanks.

Digital topographic map layers (scale: 1:10,000) of the study area were obtained from the Survey Department, and all GIS (shape) files were corrected with on-screen editing facilities available with Quantum GIS 2.8.8 free software with online base maps (Google Earth Global Viewer). Draft-corrected VTCS maps were finalized after comprehensive field verification with

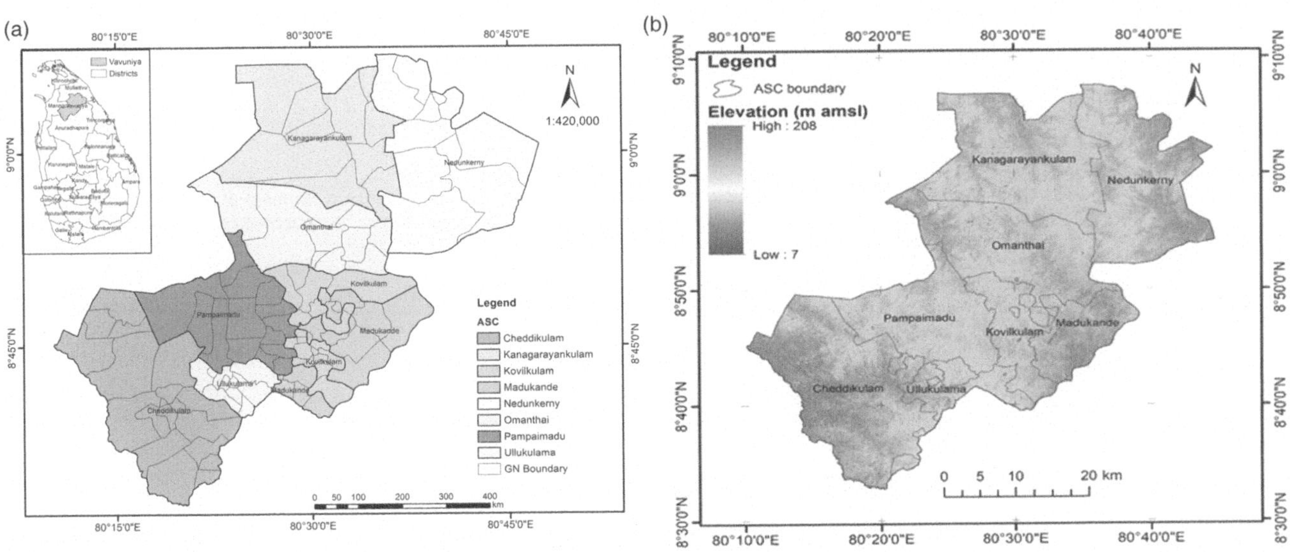

Figure 1 | Study area: (a) Eight ASCs of Vavuniya district and (b) Topographic map.

hand-held global positioning system (GPS) device (eTrex 10). The VTCS map was used as a guide map during all field-level investigations.

The digital elevation model (DEM) with a 100 m resolution was developed using shape file format elevation contour data with interpolation tool available with ArcGIS 10.2.2 software and used for the hydrological analysis using the same software.

The DEM of the Vavuniya district was clipped to create the study area's flow direction map. The DEM was created using contour maps in the format of contour lines and points. A triangulated irregular network (TIN) was created using elevation information and used to create the DEM. The TIN file was developed using the TIN tool in Arc GIS 10.2.2. The DEM file for the study area was created as a square that could cover the total study area. Hence, it would not collapse the stream flow.

Fill-sink, flow accumulation and flow direction maps were prepared using hydrological tools in Arc GIS 10.2.2 to understand the pattern of natural water flow or cascade system. A fill-sink map was created to get depression-free DEM surface for the error-free running of hydrological feature extraction algorithms in hydrological feature development using hydrologic tools of ArcGIS. A flow direction map was created to determine the direction in which water would flow out of each cell. A flow accumulation map was created to calculate the number of upslope cells flowing to a location and hence create a stream network. Flow lines were extracted using threshold values of 100, 200, 500 and 1,000 m. The most suitable threshold value for the study area was selected based on the field investigations, records and Google map. Finally, the created flow accumulation map was overlaid with the digitized tank shape file, and then the VTCS map was developed. Mapping accuracy was assessed using field verifications. The exact threshold value was selected as 100 m for the flow accumulation map. Village tank cascades were verified and validated at the field level using local people's knowledge in the respective village tank cascades.

RESULTS AND DISCUSSION

Based on the secondary data available at the Office of the Asst. Commissioner of Agrarian Services from 2017, there were 756 village tanks in the eight ASCs of the Vavuniya district. However, this study identified a total of 80 cascades comprising 514 village tanks using the GIS approach. Out of the identified number of cascades, 69 cascades were validated in the field using the knowledge of local people. In addition, this study identified 111 isolated village tanks (15%) without connecting with other village tanks and 131 abandoned village tanks (17%) in the district. The summary results of the above are illustrated in Table 1.

As shown in Table 1, the *Pampaimadu* ASC is comprised of a more significant number of village tanks in the Vavuniya district than that of the rest of the ASCs, while the *Kanagarayankulam* ASC has the lowest number. However, in terms of interconnected tanks, i.e., cascades, more cascades were identified in *Omanthai* ASC than that of other ASCs, and the lowest number was recorded in *Madukanda* ASC in the desk study.

Among the ASC divisions, the *Omanthai* ASC is found to have the highest variation in a number of cascades identified and validated (six). In contrast, in *Pampaimadu*, *Ulukulam* and *Kanagarayankulam* ASC divisions, the variation is only two cascades in each, and in *Madukanda* only one cascade. This may be due to the fact that there was a temporal variation in

Table 1 | Summary results of the study on the village tank cascades in Vavuniya district

ASC division						
Name	**Area (m^2)**	**Number of cascades identified in the desk study**	**Number of cascades validated in the field**	**Total number of tanks in the cascades**	**Number of isolated tanks**	**Number of abandoned tanks**
Cheddikulam	428	16	13	90	22	43
Pampaimadhu	235	16	14	124	14	28
Kovilkulam	111	05	07	82	03	05
Madukanda	137	04	03	49	08	03
Nedunkerny	336	07	10	47	07	30
Omanthai	289	17	11	68	24	10
Ulukulam	86	07	05	35	02	02
Kanagarayankulam	372	08	06	19	31	10
Total		**80**	**69**	**514**	**111**	**131**

the identification and the field validation of cascades, as the degree of stream/water flow in the cascades is highly dependent on seasons.

Based on the field validation, the highest number of cascades (14) were observed in *Pampaimadu* ASC, and the lowest number (three) of cascades were notified in *Madukanda* ASC. Therefore, in order to show the operative cascades, the authors take *Pampaimadu* ASC and *Madukanda* ASC as two examples, as these ASCs are validated with the highest and lowest number of cascades, respectively. The field study explores the *Kovilkulam* ASC with the highest number of village tanks per cascade, and *Kanagarayankulam* ASC with the least number of village tanks per cascade (Table 2).

Figure 2 shows three distinct cascades (along with tanks and water flows) in the *Madukanda* ASC which are notified as A, B and C. Field validation confirmed that 46 village tanks connected to the three cascades in the said ASC, while eight village

Table 2 | Minimum and maximum number of village tanks connected in cascades of each ASC in the Vavuniya district

Name of the ASC division	Minimum number of village tanks in a cascade	Maximum number of village tanks in a cascade
Cheddikulam	02	23
Pampaimadhu	02	23
Kovilkulam	03	47
Madukanda	05	23
Nedunkerny	02	23
Omanthai	03	11
Ulukulam	02	12
Kanagarayankulam	02	05

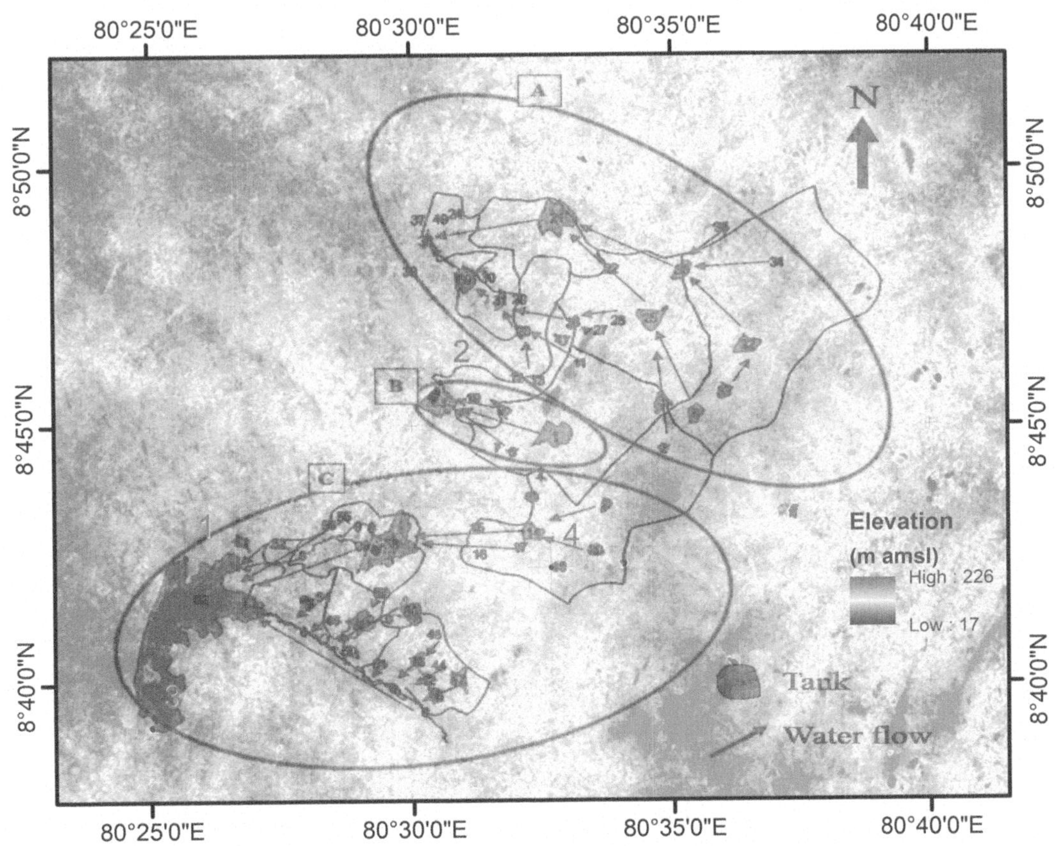

Figure 2 | Cascade validated in the Madukanda ASC division (three cascades: A–C).

tanks are isolated and three are abandoned. The isolated tank is not connected to any other tank, and there is no water flow from or into the tank.

However, in *Pampaimadu* ASC (as shown in Figure 3), the highest number of cascades, i.e., 14 distinct cascades, are observed in the field study, and 124 village tanks are connected to the existing cascades. These 14 distinct cascades are marked A–N, as shown in Figure 3. The cascades of *Pampaimadu* ASC are highly fragmented compared to *Madukanda* ASC, as a greater number of distinctive cascades are observed. The *Madukanda* ASC does not seem highly fragmented as fewer cascades are validated in the field. *Madukanda* ASC consists of cascades connecting from five to 23 village tanks per cascade, while Pampaimadu ASC ranges from two to 23 village tanks per cascade (Table 2). In total, 14 village tanks have been isolated, and 28 village tanks abandoned in the Pampaimadu ASC, which accounted for 25% of the total tanks in the said ASC. The cascade connected to a maximum number of village tanks is expected to have more water conserved (Panabokke *et al.* 2002; Matsuno *et al.* 2003).

On the contrary, the ASC divisions, namely *Kovilkulam* and *Nedunkerny*, were found, respectively, with two and three more cascade lines to be operational during the field validation, but not identified in the desk study. This shows that the connectivity of the tanks was fragmented, leaving a more significant number of small cascades in the field as operational. This could be due to tank siltation (Bebermeier *et al.* 2017), human interventions, such as encroachments, and improperly planned developmental activities (Somaratne *et al.* 2005). In addition, the improper developmental activities undertaken in the particular area would have let the water flow lines find alternative paths in line with the terrain. Somaratne *et al.* (2005) added that water shortage and the breach of tank bund due to heavy rain and lack of financial support from government/other organizations to renovate the tanks also could be the possible reasons for the village tanks to be abandoned/isolated.

It was further noticed that there are some village tanks that remained either isolated or abandoned in each ASC. The migration of farmers toward urban areas due to inadequate income from farming and the situation of unrest that existed for nearly three decades may be the reasons for a substantial number of village tanks being abandoned. The prolonged ignorance of relevant authorities on VTCS for maintenance and rehabilitation led to 17% of tanks being abandoned and 15% of tanks being isolated.

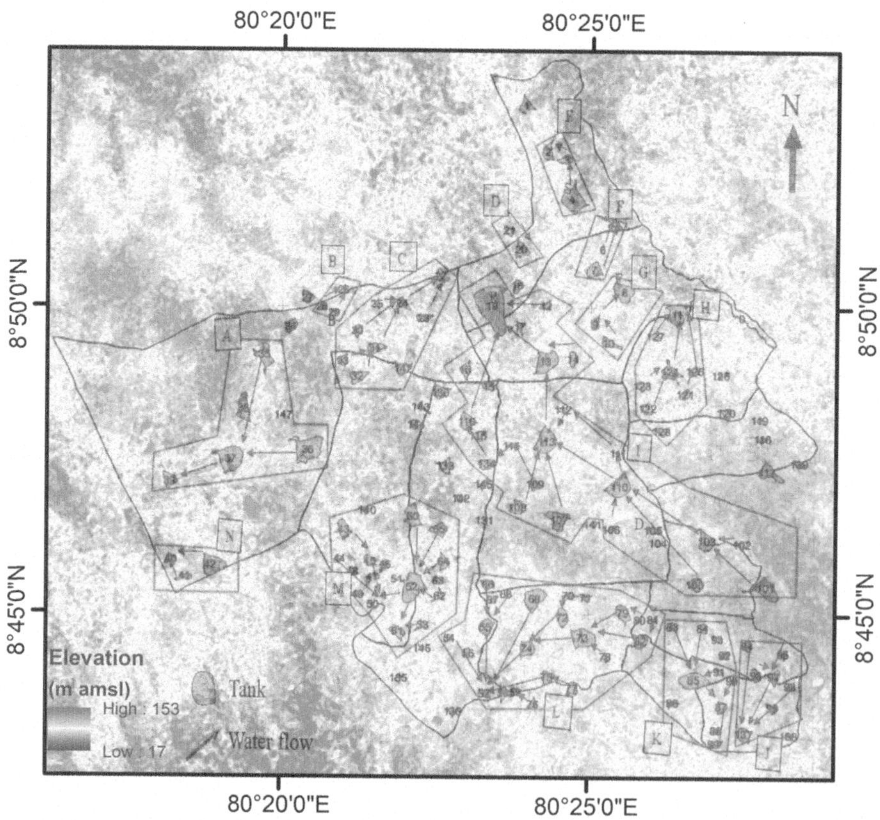

Figure 3 | Cascades validated in the Pampaimadu ASC division (14 cascades: A–N).

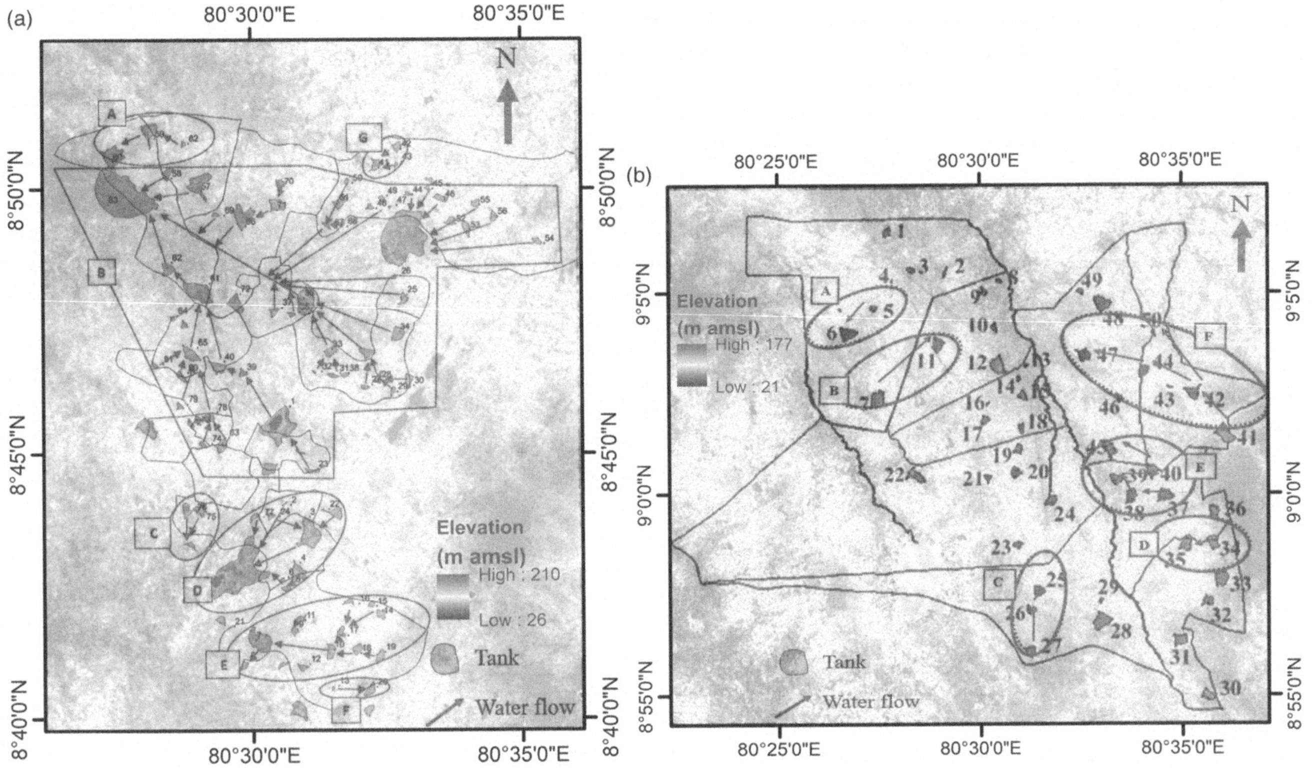

Figure 4 | Cascades validated in the field study: (a) Kovilkulam ASC and (b) Kanagarayankulam ASC.

Moreover, as shown in Table 2, the authors highlight the minimum and the maximum number of village tanks in different cascades in each ASC division in the Vavuniya district, in which the *Kovilkulam* ASC identified with a maximum of 47 village tanks in one cascade. In contrast, *Kanagarayankulam* ASC is shown with a minimum of five village tanks in a cascade (Figure 4). As the *Kovilkulam* ASC is located close to the Vavuniya town region, households continued their farming activities under these village tanks, and the desilting activities were undertaken by the farming community. Furthermore, the ASC might have obtained a source of funding for village tank rehabilitation from time to time, as the *Kovilkulam* ASC was not much affected during the civil unrest. As such, the fragmentation of cascade lines might not have happened in the same way as other ASCs in the Vavuniya district. However, this is contrary to Somaratne *et al.* (2005) and Shanmuganathan *et al.* (2010), as the disruption of cascades due to human settlements are more prone to the town area.

In *Kanagarayankulam* ASC, the number of tanks validated in cascades is 19, while the number of isolated tanks identified is 31 which is the highest number among the ASCs studied. *Kanagarayankulam* ASC is located in the north of Vavuniya and was affected severely during the civil unrest (1990–2009); during that period no farming activities were undertaken due to the mass displacement of the households. No rehabilitation or desilting was carried out for a long time which meant a higher number of village tanks were left isolated due to natural siltation on the water flow lines.

Therefore, pinpointing the primary cause for the fragmentation of cascades is still challenging, and the research should be done in-depth incorporating the historical records and interviews with key stakeholders in the relevant villages regarding water flow lines. Further investigation is recommended to study the possibilities of increasing the cascade areas in the study area by connecting isolated tanks with cascades (connecting the village tanks for water flow from one village tank to another). Initiation taken toward rehabilitation of cascades would enhance the livelihoods of farming households in the Vavuniya district and lead to sustainable water resource management.

As shown in Table 1, nearly 17% of abandoned tanks need to be restored to harvest the seasonal rainwater which could be achieved by the rehabilitation of the abandoned village tanks. As per the previous literature (Dharmasena 2009; Aheeyar 2013; Melles & Perera 2020), social and economic conditions in these regions should also be considered in addition to the engineering aspects of rehabilitation.

CONCLUSIONS

The identification and mapping of VTCS separately for each ASC in the Vavuniya district were performed through GIS approaches with available digital maps, remote sensing images, field investigation and validation in consultation with local older people in the respective village tanks.

In total, 69 VTCSs comprise 514 village tanks validated in this study that would be used for a rapidly developing region like Vavuniya district for sustainable water resources management integrating ecological aspects in an environmentally friendly manner. There have also been 111 isolated village tanks and 131 abandoned village tanks identified in the Vavuniya district. Furthermore, the initiation must be taken to connect those isolated tanks into cascade lines, thereby increasing the water availability for agriculture. Moreover, rainwater harvested in abandoned village tanks would contribute to groundwater recharge in this region. Hence, taking initiatives to rehabilitate and conserve those abandoned village tanks would be essential for sustainable water resource management and livelihood enhancement. These documented VTCSs along with water flow could be used by planners, engineers, environmentalists, hydrologists, researchers, etc., in the planning and implementation of water resources management projects in the Vavuniya district.

ACKNOWLEDGEMENTS

The authors would like to acknowledge the University of Jaffna, Sri Lanka, for financial support.

DATA AVAILABILITY STATEMENT

All relevant data are included in the paper or its Supplementary Information.

REFERENCES

Aheeyar, M. 2013 *Alternative Approaches to Small Tank/Cascade Rehabilitation: Socio-Economic and Institutional Perspective.* Research Report 162. Hector Kobbekaduwa Agrarian Research and Training Institute, Colombo, Sri Lanka.

Bebermeier, W., Meister, J., Withanachchi, C. R., Middelhaufe, I. & Schütt, B. 2017 Tank cascade systems as a sustainable measure of watershed management in South Asia. *Water* **9**, 231–247. https://doi.org/10.3390/w9030231.

Dharmasena, P. B. 2009 Sustainability of small tank irrigation systems in Sri Lanka at the 21st Century. In: *Traversing No Man's Land: Interdisciplinary Essays in Honour of Sadurawan Madhuma Bandhura* (Bandara, S. M., ed.). Godage International Publishers, Colombo, Sri Lanka, pp. 233–252.

Dharmasena, B. P. 2020 Cascaded tank-village system: present status and prospects. In: *Agricultural Research for Sustainable Food Systems in Sri Lanka* (Marambe, B., ed.). doi:10.1007/978-981-15-2152-2_3.

Galagedara, L., Mowjood, M. I. M. & Shamen, V. 2018 Village tank cascade systems as an adaptation strategy for climate change and agricultural sustainability. In: *CSBE AGM and Technical Conference*. doi:10.13140/RG.2.2.16662.98882.

Lakshman, G., Mowjood, M. I. M. & Shamen, V. 2019 Village tank cascade systems as an adaptation strategy for climate change and agricultural sustainability. In: *CSBE AGM and Technical Conference*, Canada.

Madduma Bandara, C. M. 1985 Catchment ecosystems and village tank cascades in the dry zone of Sri Lanka a time-tested system of land and water resource management. In: *Strategies for River Basin Management* (Lundqvist, J., Lohm, U. & Falkenmark, M., eds). D. Reidel Publishing Company. https://doi.org/10.1007/978-94-009-5458-8_11.

Mahatantila, K., Chandrajith, R., Jayasena, H. & Ranawanal, K. 2008 Spatial and temporal changes of hydrogeochemistry in ancient tank cascade systems in Sri Lanka: evidence for a constructed wetland. *Water and Environment Journal* **22**, 17–24.

Matsuno, Y., Tasumi, M., Hoek, W., Sakthivadivel, R. & Otsuki, K. 2003 Analysis of return flows in a tank cascade system in Sri Lanka. *Paddy and Water Environment* **1** (4), 173–181.

Melles, G. & Perera, E. D. 2020 Resilience thinking and strategies to reclaim sustainable rural livelihoods: cascade tank-village system (CTVS) in Sri Lanka. *Challenges* **11** (2), 27. https://doi.org/10.3390/challe11020027.

Nanthakumaran, A. & Palanisami, K. 2010 Assessment of the potential of groundwater supplementation by estimating the stabilization value of tank irrigation systems in Tamil Nadu, India. *Tropical Agricultural Research* **22**, 84–93.

Panabokke, C. R., Sakthivadivel, R. & Weerasinghe, A. D. 2002 *Evolution, Present Status and Issues Concerning Small Tank Systems in Sri Lanka.* International Water Management Institute, Colombo, Sri Lanka.

Ratnayake, S. S., Kumar, L., Dharmasena, P. B., Kadupitiya, H. K., Kariyawasam, C. S. & Hunter, D. 2021 Sustainability of village tank cascade systems of Sri Lanka: exploring cascade anatomy and socio-ecological nexus for ecological restoration planning. *Challenges* **12**, 24. https://doi.org/10.3390/challe12020024.

Shanmuganathan, S., Manobavan, M. & Fernando, G. W. A. R. 2010 The importance of the small-tank cascade system for the sustainable production of water in the dry zone of Sri Lanka. In: *Conference: Proceed. 9th Asia-Pacific Round Table for Sustainable Consumption and Production.*

Somaratne, P. G., Jayakody, P., Molle, F. & Jinapala, K. 2005 *Small Tank Cascade Systems in the Walawe River Basin.* Working Paper 92. International Water Management Institute. ISBN 92-9090-592-1.

Sudusinghe, S. A., Nanthakumaran, A. & Kadupitiya, H. K. 2016 Identification of cascade systems and assessment of the performance of village tanks in Kovilkulam Agrarian Centre Area. In: *7th International Conference on Sustainable Built Environment*, Kandy, Sri Lanka.

Tharani, G., Nanthakumaran, A. & Karunainathan, T. 2015 Hydrological interaction between surface water and ground water in Vavuniya District of Sri Lanka. *International Research Journal of Environmental Sciences* **4** (9), 39–44.

Withanachchi, S. S., Köpke, S., Withanachchi, C. R., Pathiranage, R. & Ploeger, A. 2014 Water resource management in dry zonal paddy cultivation in Mahaweli River Basin, Sri Lanka: an analysis of spatial and temporal climate change impacts and traditional knowledge. *Climate* **2**, 329–354.

First received 5 August 2021; accepted in revised form 5 December 2021. Available online 21 December 2021

© 2022 The Authors Water Management Addressing Societal and Climate Change Challenges doi: 10.2166/wcc.2022.369

Investigating streamflow variability and hydroclimatic teleconnections: a comparison of regulated and unregulated streamflow stations

Tarun Pant, Abinesh Ganapathy and Ankit Agarwal*

Department of Hydrology, Indian Institute of Technology Roorkee, Uttarakhand, 247667, India

*Corresponding author. E-mail: ankit.agarwal@hy.iitr.ac.in

AG, 0000-0003-0411-6791; AA, 0000-0001-8572-7046

ABSTRACT

This study evaluates the relationship between flow variability of unregulated and regulated streamflow stations and global climate indicators. Mann–Kendall and change-point analysis is applied to investigate the gradual and abrupt changes in streamflow data, followed by the investigation of multi-scale fluctuations in streamflow data using Continuous Wavelet Analysis. Linkages between streamflow and global climate indicators are examined using Cross-Wavelet and Wavelet Coherence Analysis. Results showed contrasting trend values for unregulated and regulated streamflow stations. Surprisingly, all unregulated stations experienced a significant abrupt shift in change point contrary to the regulated streamflow. Further, for unregulated stations, streamflow variability and hydroclimatic teleconnections were observed at a lower scale, indicating that variations in streamflow are more frequent and generally occur on an intra-annual to inter-annual scale. Contrary to this, regulated stations observed the streamflow variability and hydroclimatic teleconnections at a larger scale (8–10 years), indicating that all the fluctuations are smoothed out. Thus, unregulated stations cannot be used as a proxy for regulated stations in any given basin. Indeed, for better water resource planning and management, both regulated and unregulated streamflow should be investigated.

Key words: climate indices, hydroclimatic teleconnections, regulated and unregulated stations, streamflow variability, wavelet analysis

HIGHLIGHTS

- Distinct characteristics of unregulated and regulated streamflow variability and hydroclimatic teleconnections.
- Intra-annual and annual variability, i.e., 0.5 and 1 year, are prominent in unregulated in contrast with the appearance of intra-decadal scale in the regulated streamflow stations.
- Hydroclimatic teleconnections of unregulated stations are prevalent up to inter-annual scale, whereas they existed up to intra-decadal scale for regulated stations.

1. INTRODUCTION

Gradual or abrupt variation in streamflow directly affects the socio-economic condition of the particular region, and it is essential to understand its variability for appropriate management and planning of the water resources (Ehsanzadeh *et al.* 2011). The streamflow variability can be analyzed with the help of trend, stationarity, homogeneity, periodicity, and noise present in the hydrologic times series (Drissia *et al.* 2019). However, many studies showed trend analysis as one of the most widely used techniques for identifying variability in hydrological time series (Rathnayake 2019). Gudmundsson *et al.* (2019) evaluated the global change in mean and extreme streamflow for more than 30,000 streamflow sites and found the contrasting hydrological change. Sahoo & Jha (2020) examined low flow long-term trends for 14 stations of the Mahanadi river basin in India and observed an increasing trend in the upper Mahanadi River basin. Further, Birsan *et al.* (2014) found increasing winter and autumn streamflow trends and decreasing summer flow trends in Romanian streamflow. Long-term streamflow variability is attributed to anthropogenic climate change and the influence of large-scale climate drivers (Jiang *et al.* 2013). Many researchers from India (Sahu *et al.* 2012; Panda *et al.* 2013; Arfan *et al.* 2019) and other parts of the world (Rochdane *et al.* 2012; Petpongpan *et al.* 2020) have also established from their study that the changing trends of hydrological variables such as streamflow or rainfall are the causal effects of varying climatic conditions. Therefore, it is essential to study flow variability and its connections with regional climate and atmospheric circulation patterns to develop probabilistic flood and drought forecasting models in assessing hydrologic risk. This would further help in decision-making for disaster management across any particular basin.

Evaluating the impact of large-scale climate patterns on streamflow has gained attention over the years. For example, Lauro *et al.* (2019) investigated hydroclimatic variability of the Colombian Andes region and associated it with El Niño Southern Oscillation (ENSO) and observed extreme discharge in the study area during El Niño. Based on the study conducted by Burn & Hag Elnur (2002), various drivers, including the Pacific Decadal Oscillation (PDO), North Atlantic Oscillation (NAO), and North Pacific index (NP), affect runoff timing in the Mackenzie River basin in northern Canada. The impact of Indian Ocean Dipole (IOD) and ENSO on extreme streamflow in the Citarum River in Indonesia was studied by Sahu *et al.* (2012) and established the strong relationship between IOD and seasonal streamflow.

Although large-scale drivers could influence local streamflow at multiple temporal scales, only a few studies are dedicated to quantifying multi-scale flow variability. Wavelet is an emerging technique extending its application to examine the variability in any time series at different scales or frequencies. Wavelet transform has been extensively used worldwide in other research to analyze streamflow variability at multiple temporal scales (Coulibaly & Burn 2004; Labat 2010; Briciu & Mihăilă 2014). Further, this method is also used to understand the mechanisms that drive flow variability and to evaluate the teleconnections of climate drivers (Joshi *et al.* 2016; Schulte *et al.* 2016; Rathinasamy *et al.* 2019). Climate change would have detrimental effects on Asia's densely populated river basins (Immerzeel *et al.* 2010). Thus, a country like India needs to understand streamflow dynamics since major rivers serve as the primary source of water supply for irrigation, power production, industrial use, and other purposes.

The river basins of the Indian subcontinent region are very sensitive to changing climate scenarios (Panda *et al.* 2013). Previously, many studies have carried out the analysis for Indian basins in the context of global climate indices influence (like ENSO & IOD) on streamflow variability (Krishna Kumar 1999; Ashok *et al.* 2001; Gadgil *et al.* 2003; Chowdhury & Ward 2004; Maity & Nagesh Kumar 2008; Maity & Kashid 2011). It is worth noting that all of the studies mentioned above are based on streamflow variability across a whole basin. However, there are numerous regulated and unregulated streamflow stations in each basin. To understand streamflow variability at a regional or local scale, it is essential to examine the responses of different stations in any basin. Beaupré *et al.* (2020) and Growns *et al.* (2000) compared regulated and unregulated streamflow stations. However, the studies for Indian basins demonstrating the response of regulated and unregulated streamflow variability and hydroclimatic teleconnections are scarce (Yeditha *et al.* 2021). Thus, trend and wavelet analysis methods are used in this study to investigate the streamflow variability of six different unregulated streamflow stations in India and quantify the influence of precipitation and global climate indices at different time scales. Furthermore, the study had expanded to the 40 regulated streamflow stations of the same chosen basins, primarily in the context of hydroclimatic teleconnections, intending to distinguish these relationships for regulated and unregulated streamflow. Thus, the current study deals with the following objectives:

1. To investigate and quantify the influence of global climate indices on streamflow variability at six unregulated and 40 regulated streamflow stations at different time scales.
2. To identify the distinction between regulated and unregulated stations regarding streamflow variability and hydroclimatic teleconnections.

2. STUDY AREA AND DATA

Indian river basins are considered in this study to evaluate the streamflow variability at multiple time scales and to assess its link with the climate drivers. Central Water Commission has divided India into 20 river basins, including 12 primary and eight composite basins. Rivers belonging to the peninsular region of India are primarily non-perennial and mainly fed by rainwater. Because of the peninsular region's stiff and rigid lithology, there is a much lower chance of adding underground water into the surface water. So, understanding the streamflow variability in this region gives more accuracy than the Himalayan River basin, in which streamflow primarily depends on snow, precipitation, and underground water. So, by considering these facts, the present study assessed six major river basins of the peninsular region of India, including Brahmani, Cauvery, Mahanadi, Narmada, Godavari, and Subarnarekha. Six unregulated stations and 40 regulated stations are selected from the selected river basins (Figure 1). Refer to the supplementary file (Table S1 and S2) for detailed information on the streamflow stations. Streamflow data sets for the selected stations are obtained from the Water Resources Information System (WRIS)

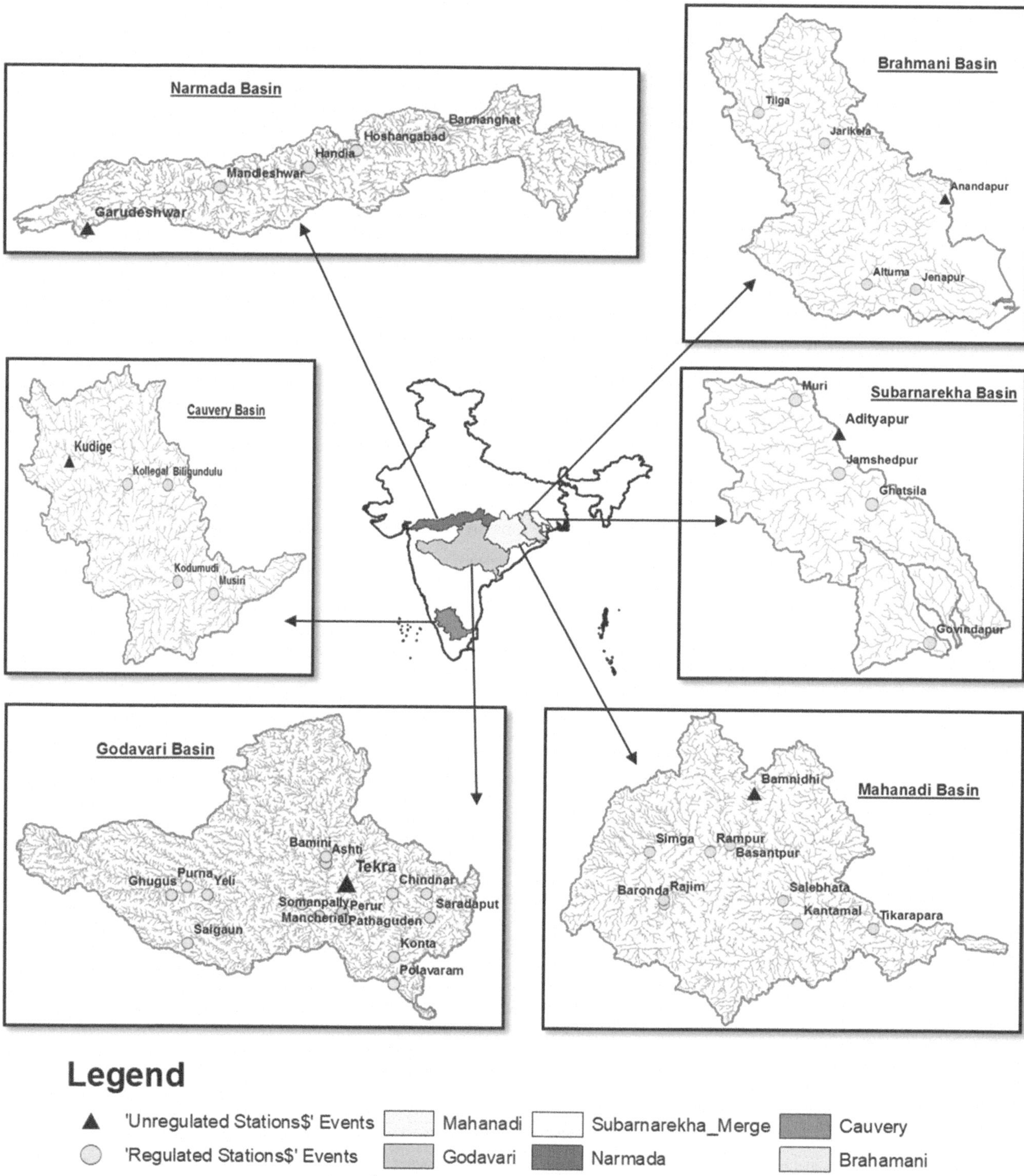

Figure 1 | Location of unregulated and regulated stations (black triangles and green dots) overlying the six river basins, under study, of peninsular Indian region.

India website http://indiawris.gov.in/wris/ for 41 years from 1975 to 2015. The daily streamflow data are converted to monthly time series using the downsample toolbox in MATLAB 2019a.

Climate indices data, namely, IOD, NAO, ENSO (Niño3.4), and PDO are used to investigate the hydroclimatic teleconnections on the regulated and unregulated streamflow station of the six Indian peninsular basins. The IOD Mode Index is calculated by subtracting the SST anomalies from the western (10°S–10°N, 50°E–70°E) and eastern (10°S–0, 90°E to 110°E) tropical Indian Oceans (Saji *et al.* 1999). The NAO is calculated based on the sea surface temperature between the North Atlantic Ocean subpolar low and high subtropical regions. The Niño3.4 index, which represents the ENSO and has better correspondence with the Indian monsoon rainfall, is computed by averaging the sea surface temperatures (SSTs) anomalies over the Niño 3.4 region (5°N–5°S, 120°W–170°W). Pacific Decadal Oscillations (PDO) is a climate index that occurs on a larger scale, with patterns primarily observed in the North Pacific Ocean. Its phases are classified as warm or cool based on temperature changes. It was often observed that large-scale patterns of atmospheric circulation like IOD (Guan & Yamagata 2003; Behera *et al.* 2006), NAO (Hurrell 2001; Visbeck *et al.* 2001), ENSO (Neelin *et al.* 1998; McLaughlin *et al.* 2014), and PDO (Mantua & Hare 2002; D'Arrigo & Wilson 2006) determine the spatio-temporal distributions of crucial atmospheric variables like surface temperature and precipitation over the land surface, which in turn control the hydrologic cycle. For this study, global climate indices data are downloaded from the website https://www.esrl.noaa.gov/psd/data/climateindices/list/

3. METHODS AND METHODOLOGY

Streamflow and global climate indices are the variables investigated in this study. Streamflow provides information on the overall response of the river basin, and climate indices influence streamflow variability. The temporal analysis of streamflow will be more concentrated in this study; however, it is necessary to understand the gradual and abrupt changes before understanding the temporal variation. So before proceeding straight to the temporal analysis trend, step-change analysis was performed to evaluate the significant gradual and abrupt changes in streamflow. Wavelet transforms due to its potential competence in evaluating the temporal variability at a varying frequency of time series signal. It is employed in this study for temporal analysis at multiple time scale and identifying hydroclimatic teleconnections. The flowchart in Figure 2 depicts the step-by-step methodology used in this study to examine streamflow variability and quantify the influence of hydroclimatic teleconnections across six river basins in India. Here, in this section, the brief methodology adopted in the study is discussed, and then the different methods used for research are explained.

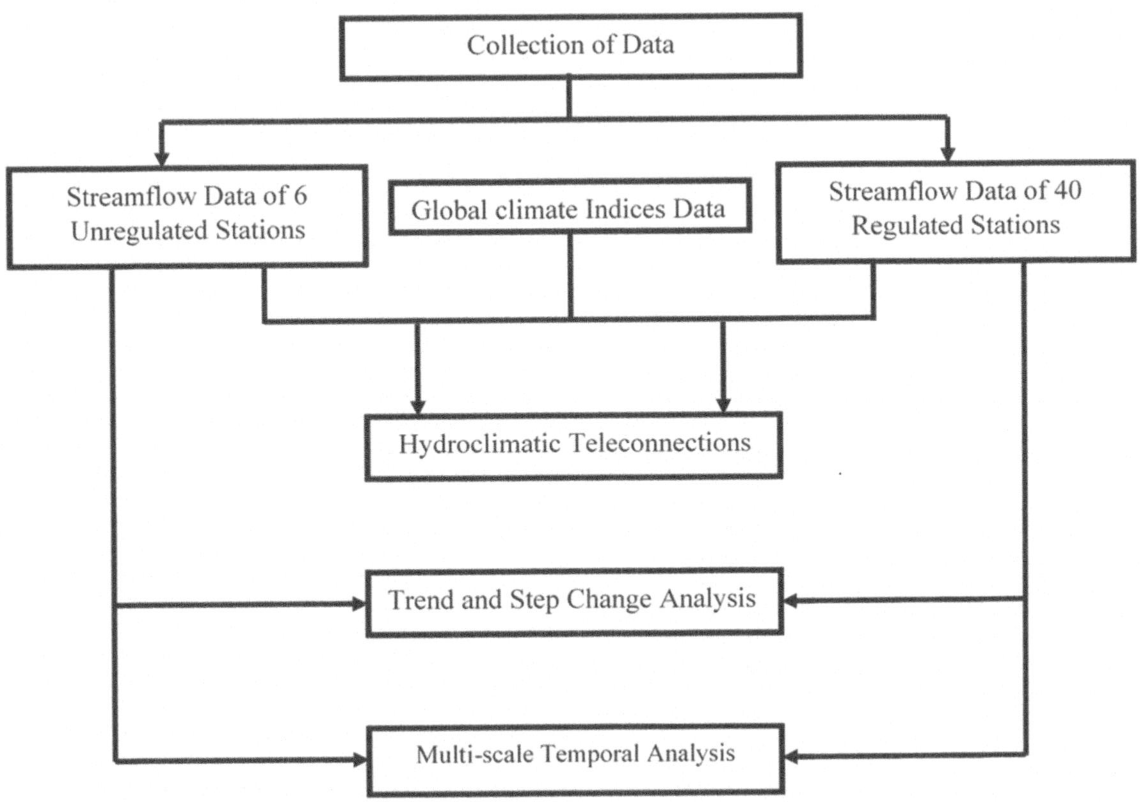

Figure 2 | The methodology adopted for the present study. Temporal variation of the streamflow data of both the regulated and unregulated stations is analysed using trend, step change, and multi-scale temporal analysis followed by evaluating its hydroclimatic teleconnections.

(1) Daily streamflow data of 41 years from 1975 to 2015 of six unregulated and 40 regulated stations in India are obtained from the Center Water Commission (CWC) Board of India. For the same period, data of four selected climate indices (IOD, NAO, Niño3.4, & PDO) are also collected.
(2) Variability in both unregulated and regulated streamflow is checked in terms of gradual and abrupt change. For finding gradual or monotonic change in streamflow data, the non-parametric Mann–Kendall trend test is used, while for abrupt or step change in streamflow data, change-point analysis is performed. MATLAB has been used for trend and step-change analysis.
(3) For further investigation, the multi-scale variability in unregulated and regulated streamflow, wavelet analysis is used. Continuous wavelet analysis is performed to identify the multi-scale variability in streamflow data.
(4) Cross-wavelet and wavelet coherence analyses are employed to identify the coherency and correlation of streamflow with precipitation and global climate indices.

3.1. Mann–Kendall trend test

To determine the monotonic or the gradual trend in unregulated streamflow series for each gauging station, the Mann–Kendall (MK) test is used. The MK test capabilities were robust in detecting trends. Many researchers worldwide in their study used the MK test for finding the trend in many domains, including the trend in hydrologic data (Burn & Hag Elnur 2002; Hamed 2009), temperature data (Alhaji *et al.* 2018), water quality parameter (Kisi & Ay 2014), meteorological variables (Gocic & Trajkovic 2013), evapotranspiration data (Shadmani *et al.* 2012), and precipitation data (Ahmad *et al.* 2015; Guntu *et al.* 2020). The Mann–Kendall test is a statistical test that is frequently applied for climate and hydrology trend analysis. The use of this test has two benefits. First, it is a non-parametric test and does not require the normal distribution of the data. Secondly, due to inhomogeneous time series, this test is sensitive to abrupt breaks. The data are evaluated as a sequence of times. The following data values are compared with each data value in time series (Alhaji *et al.* 2018).

$X_1, X_2, X_{3,\ldots\ldots\ldots}, X_n$ represent n data points where X_k represents the data point at time k. Then the Mann–Kendall statistic (S) is given by

$$S = \sum_{i=1}^{n-1} \sum_{k=i+1}^{n} sgn(X_k - X_i) \tag{1}$$

where

$$\begin{aligned} &sgn(X_k - X_i) > 1, \text{ if } X_k - X_i > 0 \\ &sgn(X_k - X_i) = 0, \text{ if } X_k - X_i = 0 \\ &sgn(X_k - X_i) < -1, \text{ if } X_k - X_i < 0 \end{aligned} \tag{2}$$

A very high positive value of S indicates an upward trend, while a low negative value indicates a downward trend. To statistically quantify the significance of the trend, the probability associated with S and the sample size, n, must be computed.

The probability associated with the Mann–Kendall statistics can be calculated from the following equation.

$$f(z) = \frac{1}{\sqrt{2\pi}} e^{\frac{-z^2}{2}} \tag{3}$$

where Z is the normalized test statistic and can be computed by

$$\begin{aligned} &Z = \left\{ \frac{S-1}{\sqrt{VAR(S)}} \right\}, \text{ if } S > 0 \\ &Z = 0, \text{ if } S = 0 \\ &Z = \left\{ \frac{S+1}{\sqrt{VAR(S)}} \right\}, \text{ if } S < 0 \end{aligned} \tag{4}$$

where,

$$Var(S) = \frac{1}{18}\left[n(n-1)(2n+5) - \sum_{k=1}^{q} t_k(t_k-1)(2t_k+5)\right] \quad (5)$$

Based on the value of standardized statistic (Z), the p-value (Y) is calculated, which can be mathematically represented by the following equation.

$$Y = 2(1 - nrmcdf(Z)) \quad (6)$$

where Y is the p-value, *nrmcdf* denotes the normalized cumulative distribution function of standardized statistic. The level of significance is indicated by α and if the value of $Z > Z_{1-\frac{\alpha}{2}}(=1.96)$, it determined that rejection of the null hypothesis indicated the presence of a trend, and if not, it indicated acceptance of the null hypothesis, absent of a trend. In this work, the Mann–Kendall test is carried out using software MATLAB2019a.

3.2. Step-change analysis

Change points are abrupt variations in time series data. Such abrupt changes could represent transitions between states. The detection of change points is helpful in time series modelling and prediction. Time series data are time series measurements that describe the behaviour of systems. These behaviours may change over time due to external influences, events, or systematic internal distribution changes (Kawahara & Sugiyama 2012). Change-point detection (CPD) is the problem of finding abrupt changes in data when a property of the time series changes (Reeves *et al.* 2007). Along with hydrologic data, the CPD method has been studied for a variety of purposes over the last several decades, for example, medical condition monitoring (Yang *et al.* 2006), speech recognition (Rybach *et al.* 2009), image analysis (Radke *et al.* 2005), human activity analysis (Tran 2019), and climate change detection (Ducré-Robitaille *et al.* 2003; Reeves *et al.* 2007; Itoh & Kurths 2010). In this study, a Changepoint analyzer provided by Taylor (2000) and software MATLAB2019a were used to find the abrupt changes in unregulated streamflow data.

3.3. Continuous wavelet transform

Because hydrologic data are non-stationary, their frequency content varies over time. Time-frequency decomposition is one of the methods for obtaining changing frequency information in any time series. Various methods for time-frequency analysis of non-stationary signals exist. The popular short-time Fourier transform (STFT) method generates a time-frequency spectrum by performing the Fourier transform over a specified time window (Cohen 1995). But, in STFT, the time-frequency resolution is constant, which is not suitable for non-stationary data, whose parameters are changing over time. Over the past two decades, the wavelet transform has been applied in many branches of science and engineering. Unlike the STFT, which has a constant resolution at all times and frequencies, the wavelet transform has a good time and poor frequency resolution at high frequencies and good frequency and insufficient time resolution at low frequencies (Daubechies 1990).

The continuous wavelet transform (CWT) is a part of the wavelet analysis method, primarily used to display and analyze signal characteristics that vary with time and scale. As a result, it can be a valuable tool for detecting and identifying signals with unusual spectral characteristics, transient information content, or other non-stationary properties (Sadowsky 1996). At both time and frequency resolutions, the CWT provides helpful information about any time series.

Mathematically, CWT is defined as the sum overall time of the signal multiplied by scaled, shifted versions of the wavelet function ψ. The coefficients of the wavelet transform of continuous signal $f(t)$ are defined by a linear integral operator (Maheswaran & Khosa 2012).

$$W(a, \tau) = \int_{\infty}^{-\infty} f(t)\psi_{a,\tau}(t)dt \quad (7)$$

$$\psi_{a,\tau}(t) = \frac{1}{\sqrt{a}}\psi\left(\frac{t-\tau}{a}\right) \quad (8)$$

The function $\psi(t)$, which can be natural or complex, plays the role of a convolution kernel and is called a wavelet. The parameter a can be interpreted as a dilation ($a>1$) or contraction ($a<1$) factor of the wavelet function $\psi(t)$ corresponding to different scales of observation. The parameter 'τ' can be interpreted as a temporal translation or shift of the function $\psi(t)$ which allows the study of the signal $f(t)$ locally around the time 'τ'.

The wavelet function $\psi(t)$ has the following properties.

(1) The function integrates to zero:

$$\int_{-\infty}^{\infty} \psi(t)dt = 0 \tag{9}$$

(2) The function is square integral or, equivalently, has finite energy

$$\int_{-\infty}^{\infty} |\psi(t)|^2 dt < \infty \tag{10}$$

The wavelet transforms $W(a, \tau)$ are also called *wavelet coefficients* in which the dilation parameter is equivalent to the window size in the windowed Fourier transform. By changing the dilation parameter, the information contained in different data frequencies can be analyzed independently. A small dilation value may be used to analyze high-frequency information, while a more significant value may be used to analyze a process with low-frequency components. The wavelet transform, as defined by Equation (7), is called the continuous wavelet transform (abbreviated CWT) because the scale and time parameters, a and τ, assume continuous values. It provides a redundant representation of a signal as the CWT of a function at scale 'a' and location'τ' can be obtained from the CWT of the same function at other scales and locations. Since the CWT behaves like orthonormal basis decomposition, it can be shown that it is also isometric as it preserves the overall energy content of the signal and, thereby, allows recovery of the function $f(t)$ from its transform by using the following reconstruction formula:

$$f(t) = \frac{1}{C_\psi} \int_{-\infty}^{\infty} \int_{0}^{\infty} a^{-2} W(a, \tau)\psi_{a,\tau}(t)da\, d\tau \tag{11}$$

where C_ψ is a constant and depends on the choice of the wavelet. The above equation suggests that the function $f(t)$ may be seen as a superposition of signals at different scales and obtained by varying the scale parameter 'a.'

3.4. Cross-wavelet transform

Cross-wavelet transform (XWT) explains the coherent and the phase relationships between two different time series. In XWT, the dominant frequency or the lowest frequency may affect the two different time series' coherency. Suppose two discrete time series X ($n=1\ldots N$) and Y ($n=1\ldots N$), the XWT can be calculated as

$$W^{XY}(a) = W^X(a) \times W^{Y*}(a) \tag{12}$$

Here $W^x(a)$ is the CWT of the time series X considered. The complex conjugate of $W^X(a)$ the CWT of time series Y. Cross-wavelet constructs the relationship between two CWTs with high standard power (covariance) of the two time series (Grinsted *et al.* 2004). The significant covariance of these time series can be observed as red noise in the wavelet scalograms.

3.5. Wavelet coherence analysis

Cross-wavelet transform sometimes may not give precise results in the case of coherency. To overcome this problem, wavelet coherence is used, which provides accurate results about coherent relationships. Wavelet coherence exhibits the correlation of two time series, which is based on time-frequency analysis. Wavelet coherence aids in understanding the relationship between low power frequencies, which cannot be correctly analyzed using a cross-wavelet. Consider two time series y and

x; then the wavelet coherence is given by

$$R(y,x) = \frac{\varsigma[W(y,x)]}{\sqrt{\varsigma[W(y)]\varsigma[W(x)]}} \tag{13}$$

$$R^2(y,x) = R(y,x).R(y,x)^*; \tag{14}$$

In the above equation,

$R\ (y, x)$ – wavelet coherence between y and x; $R^2(y, x)$ – squared wavelet coherence between y and x; $W\ (y, x)$ denotes the corresponding cross-wavelet transforms; ςdenotes smoothing factor.

Wavelet coherence ranges from 0 to 1. If the correlation coefficient value is closer to 1, there is more correlation between the two time series and vice versa.

4. RESULTS

The time-frequency analysis of streamflow data from both types of stations was used to identify streamflow variability of unregulated and regulated stations in terms of monotonic and abrupt change. Finally, four climate indicators are used to explore hydroclimatic teleconnections of streamflow. Using the Mann–Kendall trend test, only two unregulated stations from the Cauvery and Narmada basins exhibited a significant long-term negative trend. In contrast, most regulated stations in the Brahmani, Cauvery, and Godavari river basins showed a significant negative trend, while Mahanadi, Narmada, and Subarnarekha river basins observed a positive trend. Significant abrupt variations in regulated streamflow are rare; nonetheless, all unregulated stations observed significant abrupt changes in their streamflow data.

All unregulated streamflow stations observed strong features at the intra-annual (0.5 years) and annual (1 year) time scales. However, most regulated stations are dominated by oscillations from the annual to the intra-decadal scale rather than intra-annual variability. The cross-wavelet and wavelet coherence plots between streamflow and several climate indices demonstrate that the Niño3.4 index has the strongest feature at lower and higher frequencies. The association between unregulated streamflow and all global climate indices was observed up to inter-annual scale for the unregulated station, but it was observed up to intra-decadal scale for the regulated station.

The following sections provide a complete description of all of the findings.

4.1. Long-term trend and CPD for unregulated and regulated stations

The gradual and abrupt changes in streamflow data for six unregulated and 40 regulated stations from 1975 to 2015 are estimated using MK trend analysis and step-change analysis. Table 1 presents the Z value, p-value obtained from the MK test, and step-change year from the step-change detection technique. From the values tabulated, only two unregulated stations belonging to the Cauvery and Narmada basins show a significant long-term negative trend at the 5% significance level. Three out of four stations in Brahmani and Cauvery basins and 12 out of 16 stations in the Godavari river basin show a significant negative trend in regulated streamflow stations. Further, three and two stations out of eight in the Mahanadi basin show positive and negative trends, respectively. In addition, all the regulated streamflow stations in Narmada and Subarnarekha basins display positive trends.

Step-change analysis reveals the significant negative abrupt change for the unregulated stations of Brahmani, Cauvery, Mahanadi, and Narmada basins in 2015, 1985, 2007, and 2006, respectively. In addition, a positive abrupt shift is observed for the Godavari and Subarnarekha river basins in 1992 and 1985. Surprisingly, significant abrupt changes in regulated streamflow stations are much less common. Even the shifts observed are all negative except the regulated streamflow stations in the Subarnarekha basin. Interestingly, an abrupt shift is observed in the mid-1990s for most of the stations in Brahmani and Godavari basins. One out of four stations in Cauvery (CMU) and one out of eight stations in the Mahanadi (MTI) basin experienced an abrupt negative shift. Finally, no regulated stations in the Narmada basin show any significant shifts.

4.2. Multi-scale temporal analysis

Daily streamflow data from six unregulated stations and 40 regulated stations are decomposed using continuous wavelet transforms (CWT) to determine the scale-specific dominant parameters which might cause the trend and step change in streamflow data. This is followed by cross-wavelet and wavelet coherence analysis determining the hydroclimatic teleconnections of four global climate indices.

Table 1 | Results of trend analysis and step change of streamflow time series for the chosen gauging stations of Indian river basins

River	Station	Indicated as	*Z* value	*P*-value	Significance	Step change
Unregulated stations						
Brahamani	Anandapur	BA	−0.21	0.83	–	2015–
Cauvery	Kudige	CK	−3.95	7.69×10^{-5}	**	1985–
Godavari	Tekra	GT	−1.673	0.09	–	1992+
Mahanadi	Bamnidhi	MB	0.348	0.72	–	2007–
Narmada	Garudeshwar	NG	−2.909	0.01	**	2006–
Subarnarekha	Adityapur	SA	0.864	0.38	–	1985+
Regulated stations						
Brahmani	Altuma	BAL	−4.95	7.05×10^{-7}	**	1994–
	Jarikela	BJA	−1.39	0.16	–	–
	Jenapur	BJE	−10.75	0.10	**	1994–
	Tilga	BTI	−52.89	0.10	**	1994–
Cauvery	Biligundulu	CBI	−5.06	4.14×10^{-7}	**	–
	Kodumudi	CKD	−2.98	0.002	**	–
	Kollegal	CKO	−6.64	3.08×10^{-11}	**	–
	Musiri	CMU	−26.85	0.01	–	2012–
Godavari	Ashti	GAS	0.96	0.33	–	–
	Bamini	GBA	−31.51	0.11	**	–
	Chindnar	GCH	−82.87	0.01	**	–
	Ghugus	GGH	−63.12	0.02	**	1996–
	Jagdalpur	GJA	−43.22	0.11	**	1996–
	Konta	GKO	0.10	1.64	–	–
	Mancherial	GMA	−32.97	0.08	**	1991–
	Nowrangpur	GNO	50.4698	0.01	**	1996–
	Pathaguden	GPA	−8.57	0.11	**	–
	Perur	GPE	−88.70	0.10	**	1994–
	Polavaram	GPO	−8.07	6.66×10^{-16}	**	–
	Purna	GPU	6.86	6.52×10^{-12}	*	–
	Saradaput	GSA	10.22	0.14	*	–
	Saigaun	GSI	−7.07	1.48×10^{-12}	**	–
	Somanpally	GSO	−36.22	0.01	**	–
	Yeli	GYE	−46.42	0.02	**	2005–
Mahanadi	Baronda	MBA	18.37	0.11	*	–
	Basantpur	MBS	18.37	0.11	*	–
	Kantamal	MKA	37.29	0.04	*	–
	Rajim	MRA	−5.04	0.01	**	–
	Rampur	MRM	−0.41	0.67	–	–
	Salebhata	MSA	0.79	0.42	–	–
	Simga	MSI	−16.51	0.21	–	–
	Tikarapara	MTI	−23.39	0.01	**	1994–
Narmada	Barmanghat	NBA	17.33	0.04	*	–
	Handia	NHA	32.52	0.01	*	–
	Hoshangabad	NHO	30.04	0.02	*	–
	Mandleshwar	NMA	45.44	0.11	*	–
Subarnarekha	Ghatsila	SGH	30.20	0.04	*	–
	Govindapur	SGO	83.27	0.12	*	1992+
	Jamshedpur	SJA	14.74	0.10	*	1990+
	Muri	SMU	59.26	0.01	*	–

Note: '–'represents no significant trend, '*' represents a significant positive trend, and '**' represents a significant negative trend. Step change denotes the year in which change has occurred, and '–' indicates a negative, and '+' indicates a positive shift.

The CWT was used to decompose unregulated streamflow data from six stations. Figure 3 depicts the scalograms obtained from the continuous wavelet analysis. All unregulated stations studied (BA, CK, GT, NG, MB, and SA) had high intra-annual, annual, and inter-annual variability (0.5, 1, 2, and 4 years). Significant intra-annual oscillations of 0.5 years were observed primarily for stations BA, GT, NG, and SA. However, at this scale, such oscillations were absent at stations CK and MB. Except for stations CK and MB, which had some discontinuity from 1985 to 2000, annual fluctuations were active for all stations during the study period. Significant oscillations on an inter-annual scale (2 and 4 years) were observed at stations BA, MB, NG, and SA, primarily between 1990 and 2000.

Similar to unregulated stations, scalograms are developed for 40 regulated streamflow stations. All the scalograms of regulated stations are included in the supplementary section (Fig. S1) in the same order as listed in Table 1 for brevity purposes. All four stations except station BTI, from the Brahmani basin, observed high variability in streamflow from 0.5 to 8 years of scale. These oscillations are mostly predominant from 1985 to 2000 in stations BAL and BJE. Whereas, for the station, BJA streamflow variation can be observed throughout the study period. In the case of the Cauvery basin, all stations showed significant oscillation at the annual (1 year) scale. These oscillations have appeared in three ranges from 1975 to 1985, 1990 to 2000, and 2005 to 2010. Both intra-annual and inter-annual fluctuations are absent in the case of the Cauvery basin; however, some stations observed intra-decadal oscillation of 8 years of scale in the year 2000–2005. In the case of the Godavari basin, all 16 stations showed annual oscillation in streamflow; however, some stations also observed inter-annual and intra-decadal oscillation. Significant oscillation at intra-annual and annual scales can be observed for the Mahanadi basin. However, most of the stations also observed significant oscillation at inter-annual scale (up to 8 years). The station from Narmada and Subarnarekha basin observed these oscillations up to inter-annual scale, in which the Subarnarekha basin observed it up to 8 years of scale. In contrast, oscillation for the Narmada basin is restricted at 4 years of scale.

4.3. Hydroclimatic teleconnections with unregulated and regulated streamflow

4.3.1. Indian ocean dipole

The XWT plot (Fig. S2 and Figure 4) revealed a strong relationship between IOD and streamflow. From 1980 to 2010, all unregulated stations (BA, CK, GT, MB, NG, and SA) showed a strong yearly connection. Surprisingly, the annual correlation appears to be vanishing from 2000 to 2005 and has been nonexistent in recent years. But along with smaller scales, regulated stations from Brahmani, Cauvery, Godavari, Mahanadi, and Subarnarekha observed some linkage with IOD at the

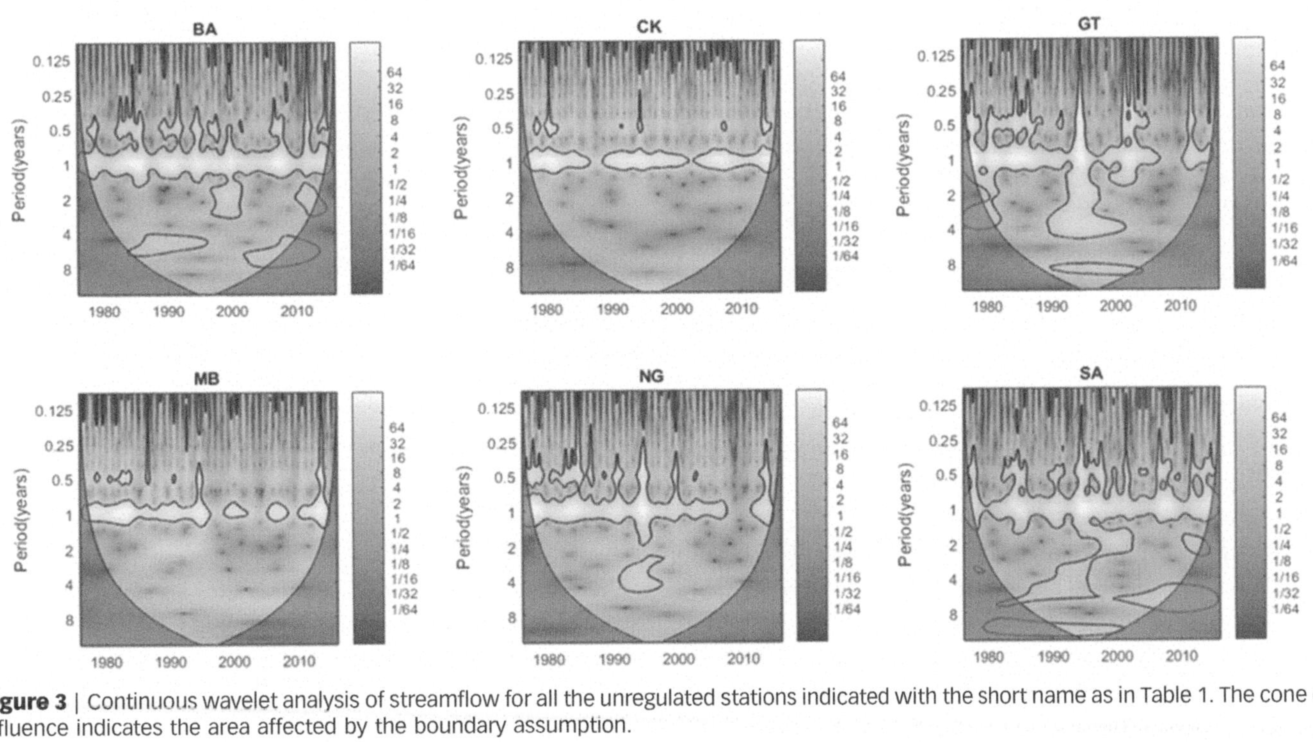

Figure 3 | Continuous wavelet analysis of streamflow for all the unregulated stations indicated with the short name as in Table 1. The cone of influence indicates the area affected by the boundary assumption.

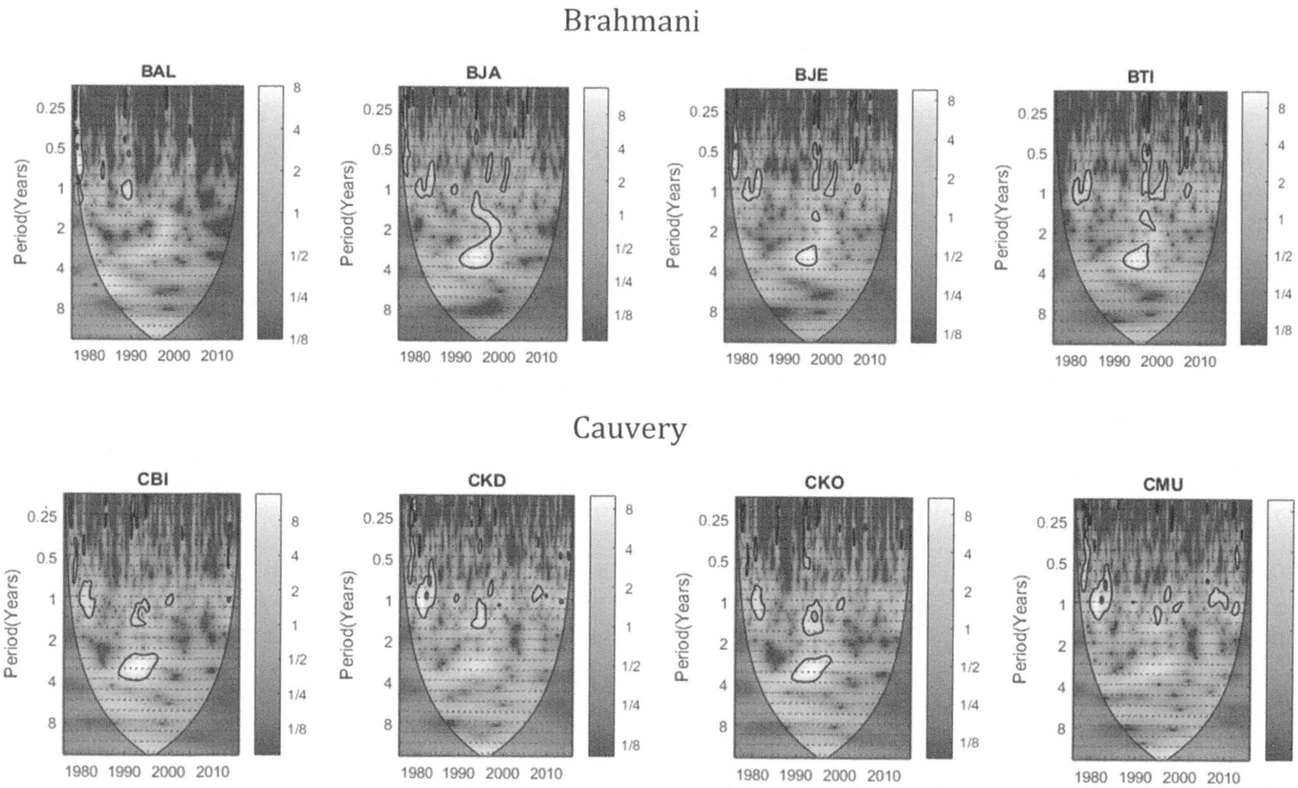

Figure 4 | Cross-wavelet analysis of IOD and 40 regulated streamflow stations of Brahmani, Cauvery, Godavari, Mahanadi, Narmada, and Subarnarekha basins in the same order listed in Table 1. The cone of influence indicates the area affected by the boundary assumption. (*continued.*).

intra-decadal scale (8 years). Mostly, the significant correlations are intermittent, but they seem to be continuous from 1985 to 2000 for most of the regulated stations. The direction of arrows in the cross-wavelet plot is not in one direction. So for a better understanding of coherence between streamflow and IOD, wavelet coherence analysis (WTC) (Fig. S3 and Figure 5) was performed. Significant areas are transient with concise duration and mainly occurred at 0.5–1, 2–4, and 4–8 years of scale. Although the IOD is an essential modulator of Indian monsoon rainfall (Ashok *et al.* 2001), its impact on streamflow variability at the regional scale is different. This research also revealed a complex link between IOD and streamflow of distinct basins at different scales and periods. From the direction of arrows, it can be observed that streamflow of the Brahmani basin has an in-phase relationship with IOD at the annual scale. However, still, it is in an anti-phase relationship at an inter-annual and intra-decadal scale. Most significant areas from Cauvery and Godavari showed an anti-phase relationship, whereas stations from Narmada and Subarnarekha observed an in-phase relationship with IOD. Mahanadi basin kept an anti-phase relationship up to 4 years of scale and an in-phase relationship with IOD at 4–8 years of scale. For the majority of the station, an anti-phase relationship can be observed from 1985 to 2000.

4.3.2. North Atlantic Oscillation

In-phase associations were observed between unregulated with NAO at the intra-annual (0.5 years) and annual (1 year) scales in XWT analysis (Fig. S4) from 1980 to 2010, with breaks in 1985, 1995, 2000, and 2005. The WTC analysis (Fig S5) produced similar results; however, stations such as GT, MB, and NG showed significant correlation at 2- and 4-year time scales.

In the case of the regulated station, except stations from the Cauvery basin, there was a significant but intermittent intra-annual correlation for all stations observed in the XWT plot (Fig. S6). Annual correlations are present for all stations of each basin, but the significant areas of high correlation are absent between the year from 1980 to 1990 at this scale. Inter-annual and intra-decadal correlation can be observed for all stations of the Brahmani and Subarnarekha basin. In contrast, only a few stations from other basins observed a significant correlation area at these particular scales, in the case of the WTC plot (Fig. S7), in which most of the stations from the Brahmani basin showed significant coherence with NAO up to 4 years of

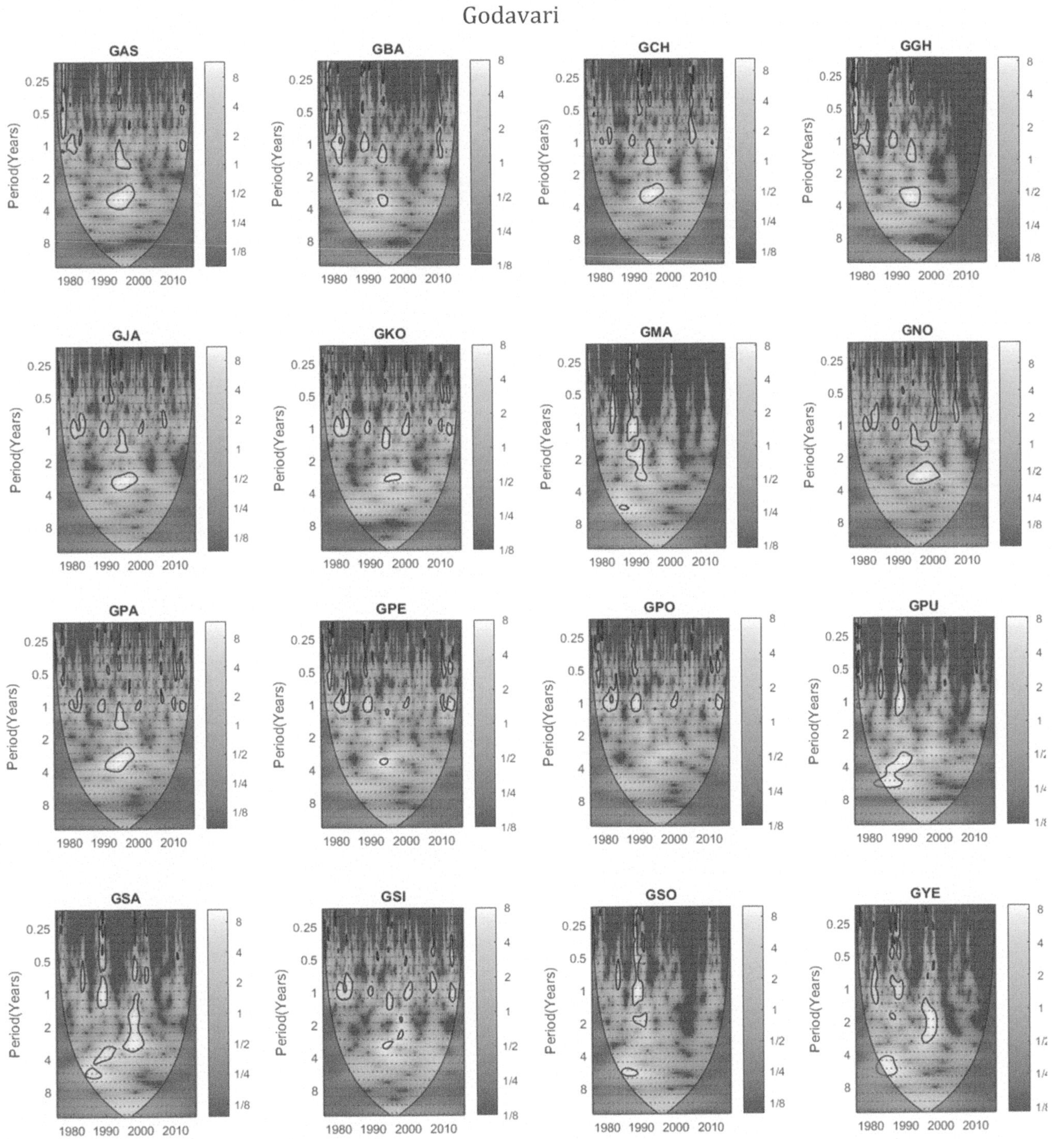

Figure 4 | Continued.

scale except for station BAL, which observed this only at the annual scale. All stations from the Cauvery basin showed almost the same coherency at intra-annual, annual, and inter-annual time scales. In contrast, stations from the Godavari basin observed distinct coherency at different time scales; for example, a station like GAS, GJA, GNO, GPA, and GPU observed significant coherency at inter decadal scale whereas, for another station, coherency can be seen only up to 4 years of scale. These similar observations can also be seen for the Mahanadi basin. The station from the Narmada basin observed the significant coherency up to 8 years of scale, whereas Subarnarekha observed this only up to 4 years of scale. In the case of the

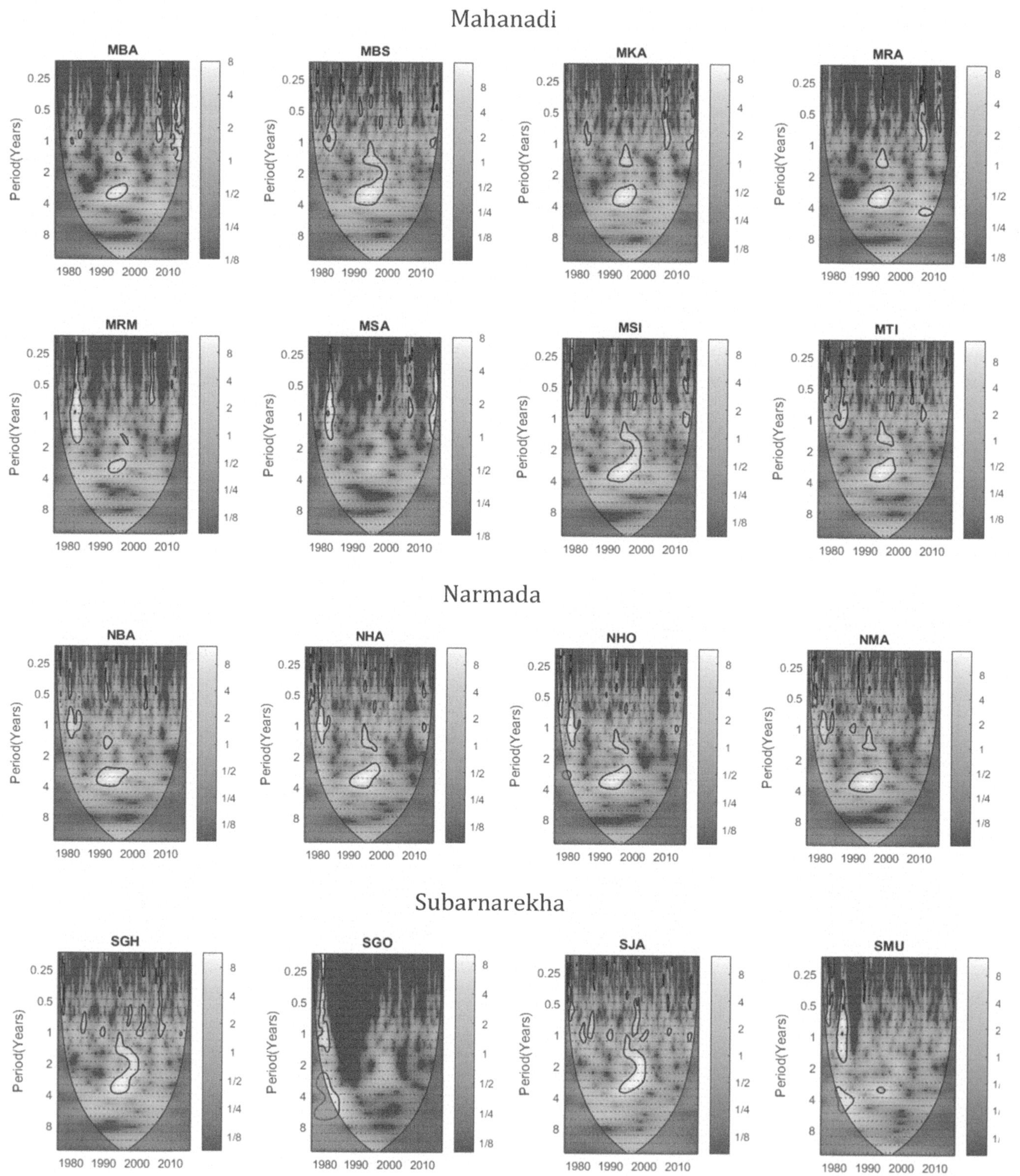

Figure 4 | Continued.

phase relationship between streamflow data and NAO, there is an in-phase relationship between Brahmani and NAO. Still, an anti-phase relationship can also be observed for these stations in 1980. The most significant area showed the upward movement of arrows for the Cauvery basin. The majority of the Godavari, Mahanadi, and Narmada stations showed an in-phase relationship, whereas Subarnarekha observed a mixed (in-phase, anti-phase, and lagged) relationship with NAO.

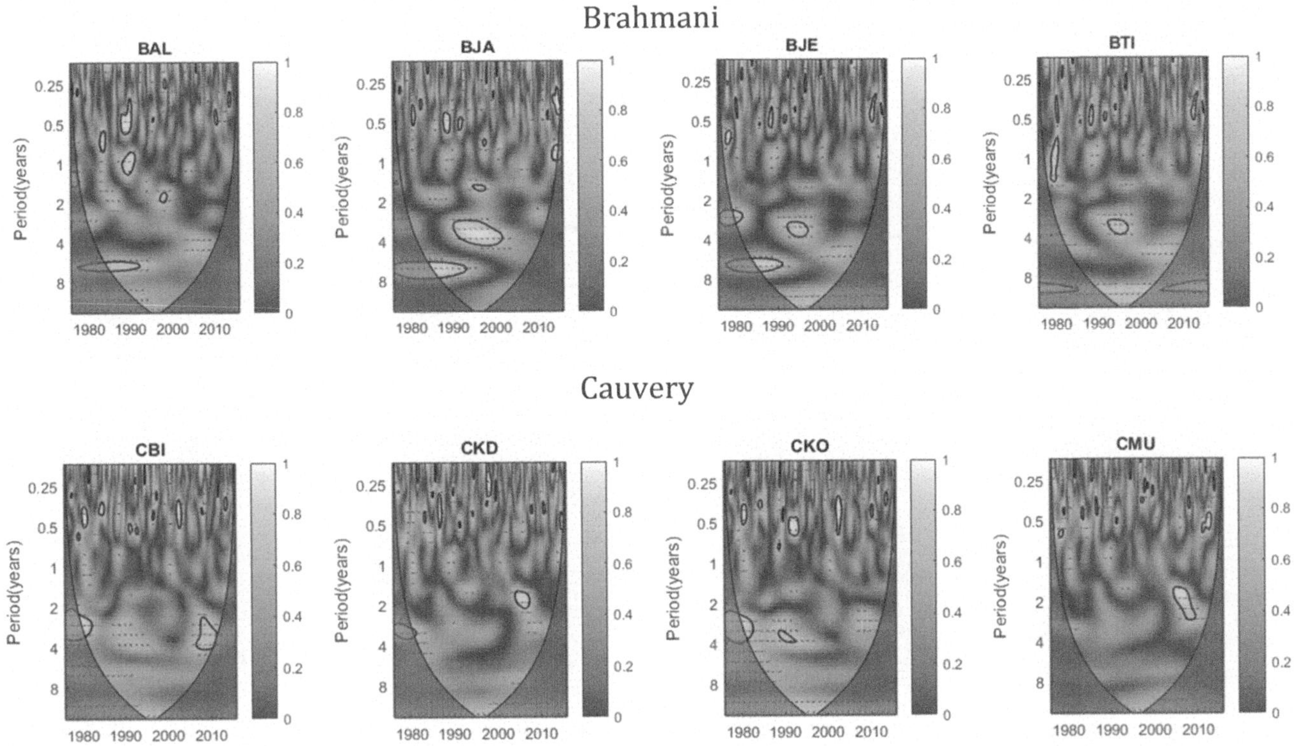

Figure 5 | Wavelet coherence analysis of IOD and 40 regulated streamflow stations of Brahmani, Cauvery, Godavari, Mahanadi, Narmada, and Subarnarekha basins in the same order listed in Table 1. The cone of influence indicates the area affected by the boundary assumption. (*continued.*).

4.3.3. Niño3.4

The study found that Niño3.4 has the strongest association with unregulated streamflow on an annual scale compared to other indices. Based on the XWT plot (Figure 6), unregulated stations such as BA, CK, MB, and SA are highly correlated with Niño3.4 throughout the period (1975–2010), whereas GT and NG have some vanishing features from 1995 to 2005. The WTC plot (Figure 7) observed periodicities at intra-annual, inter-annual, and intra-decadal scales along with the annual scale.

The XWT plot (Figure 8) shows a significantly high correlation between Niño 3.4 and streamflow stations from an annual to an intra-decadal scale for regulated stations. Annual correlations are intermittent and mostly absent in the years 1982, 1988, and 1997. Inter-annual and intra-decadal correlations are continuous for the period from 1980 to 2005. Interestingly none of the stations from any basins observe any intra-annual correlation with Niño 3.4 except the appearance of some covariance between 0.5 and 1-year scale can be observed for the entire basin in the WTC plot (Figure 9). For the Brahmani basin, a significant area with high coherency can be seen at 2–4 years of scale; however, station BJA was also observed at 4 to 8 years of scale from 1995 to 2005. Cauvery basin is not influenced by Niño 3.4 at annual, inter-annual, and decadal scale, but some intermittent significance coherency can be observed for inter-annual and intra-decadal scales. The station from the Mahanadi basin also observed some small significant areas from 0.5 to 8 years of scale.

Similarly, the Narmada basin showed coherency up to 4 years of scale, whereas Subarnarekha observed a highly significant area with high covariance at 4–8 years of scale. Overall, Niño 3.4 influences the streamflow of the chosen basin with a very short-lived significant region. These regions can be found in the years 1978, 1982, 1985, 1992, and 2006 for smaller scale, and for larger scale, these can be found in 1990–2000 and 2005. The direction of arrows is not in one direction in both XWT and WTC plots. Still, most insignificant regions showed an anti-phase relationship with Brahmani and Narmada, Leading with Cauvery, and in-phase relationship with Godavari, Mahanadi, and Subarnarekha basins.

4.3.4. Pacific decadal oscillation

PDO indices showed a leading association with unregulated streamflow at a 1-year scale in the XWT plot (Fig. S8). However, stations GT, MB, and SA observed these periodicities at 2, 4, and 8 years of scale in the WTC plot (Figure 9), in addition to 1 year of scale. These associations can be seen primarily between 1985 and 2000.

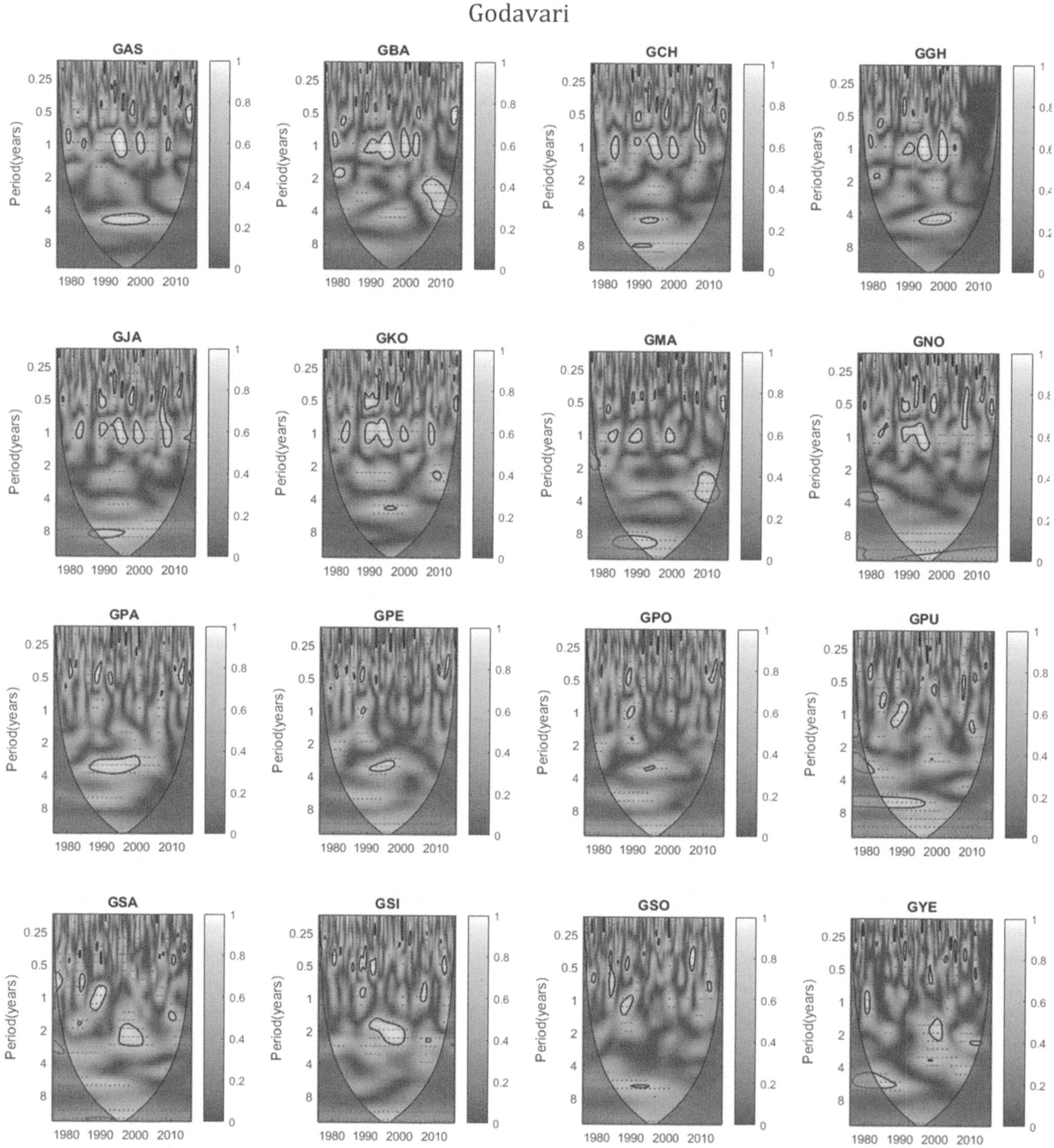

Figure 5 | Continued.

In the case of regulated stations, the XWT plot (Fig. S10) between PDO and streamflow give a similar result as of Niño 3.4 and observe the significant correlation from annual to intra-decadal scale; however, the WTC plot (Fig. S11) provides some different features of covariance with PDO, for example, Brahmani and Cauvery basin observe significant coherence at 0.5, 1, and 2–4 years of scale. All the considerable areas are short-lived and intermittent. Each station from the Cauvery basin observed similar coherence with PDO. Stations from Godavari basin and Mahanadi observed these coherencies mainly at inter-annual and intra-decadal scale, mostly from 1985 to 2000 and small coherency at 0.5 and 1 year of scale. Narmada

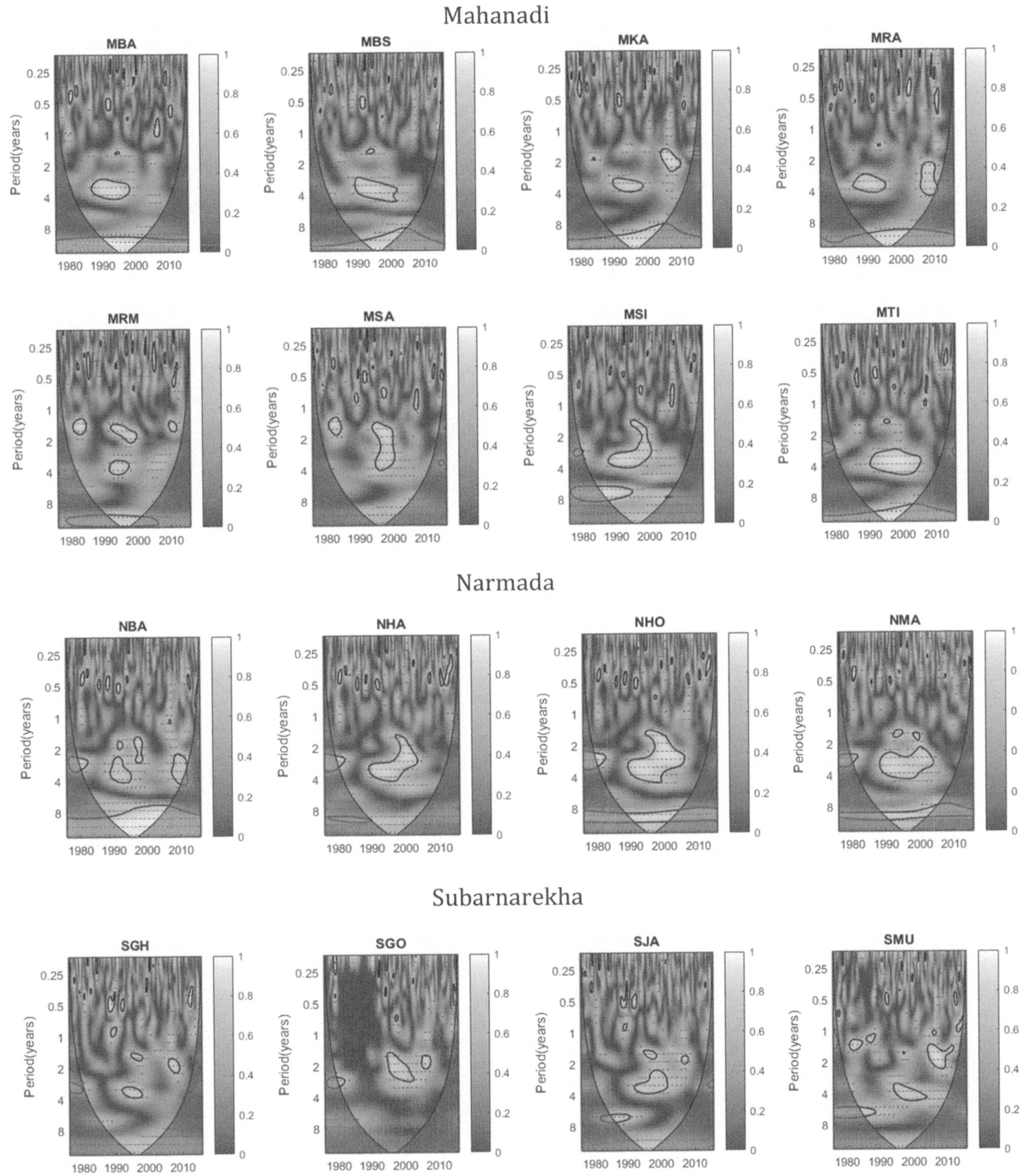

Figure 5 | Continued.

and Subarnarekha basins also observed coherency at 0.5, 1, 2, and 4 years of scale. These covariances are of short duration up to the annual scale, but from 2 to 4 years of scale, these significant regions ranged mainly from 1985 to 2000. From the direction of the arrow, it can be observed that the station belonging to Brahmani, Cauvery, Narmada, and Subarnarekha showed

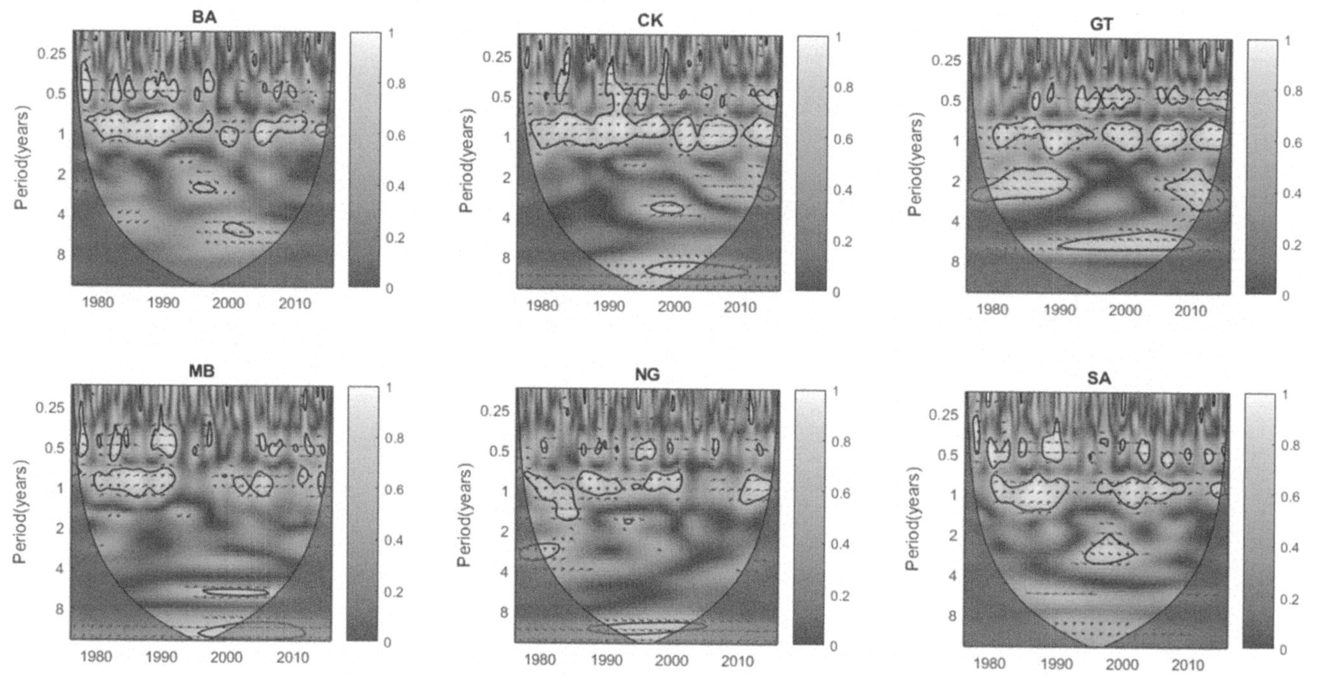

Figure 6 | Cross-wavelet analysis of Nino3.4 and unregulated streamflow stations of Brahmani, Cauvery, Godavari, Mahanadi, Narmada, and Subarnarekha basins indicated with the short name as in Table 1. The cone of influence indicates the area affected by the boundary assumption.

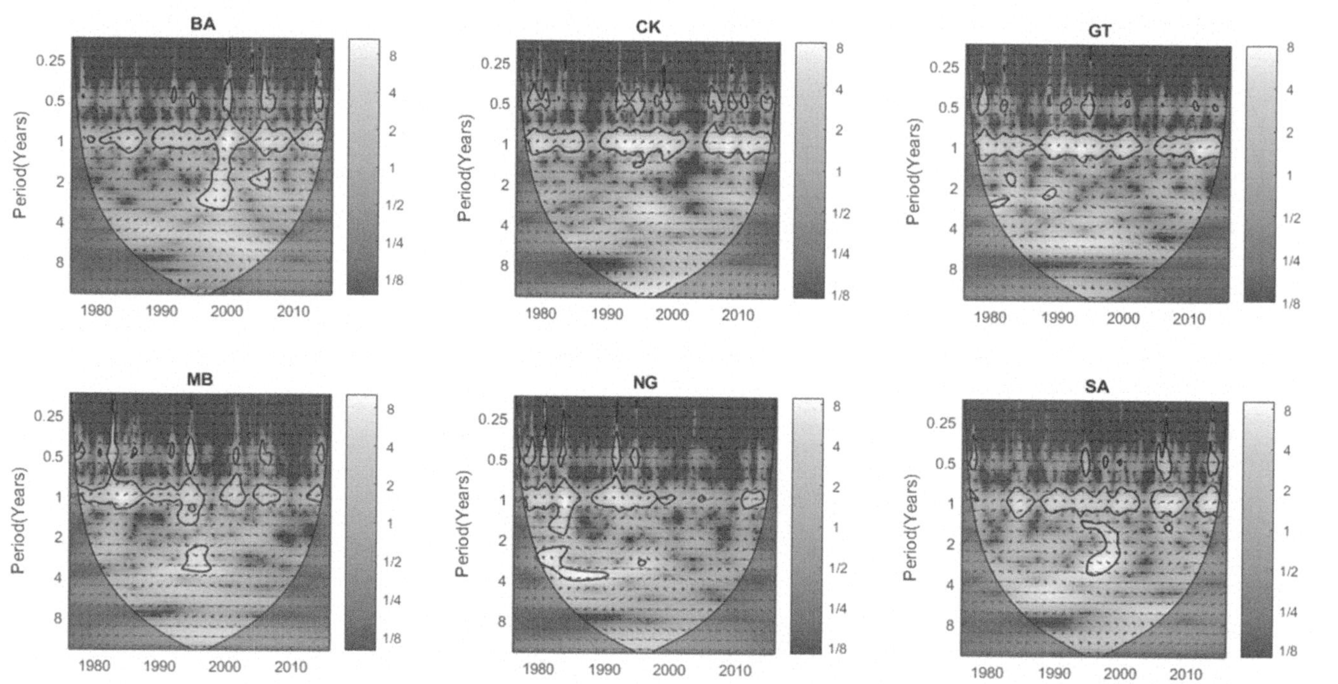

Figure 7 | Wavelet coherence analysis of Nino3.4 and unregulated streamflow stations of Brahmani, Cauvery, Godavari, Mahanadi, Narmada, and Subarnarekha basins indicated with the short name as in Table 1. The cone of influence indicates the area affected by the boundary assumption.

an in-phase relationship with PDO. In contrast, the Godavari basin observed an anti-phase relationship. The station from the Mahanadi basin showed an in-phase relationship at the inter-annual scale, whereas there was an anti-phase relationship at the intra-decadal scale.

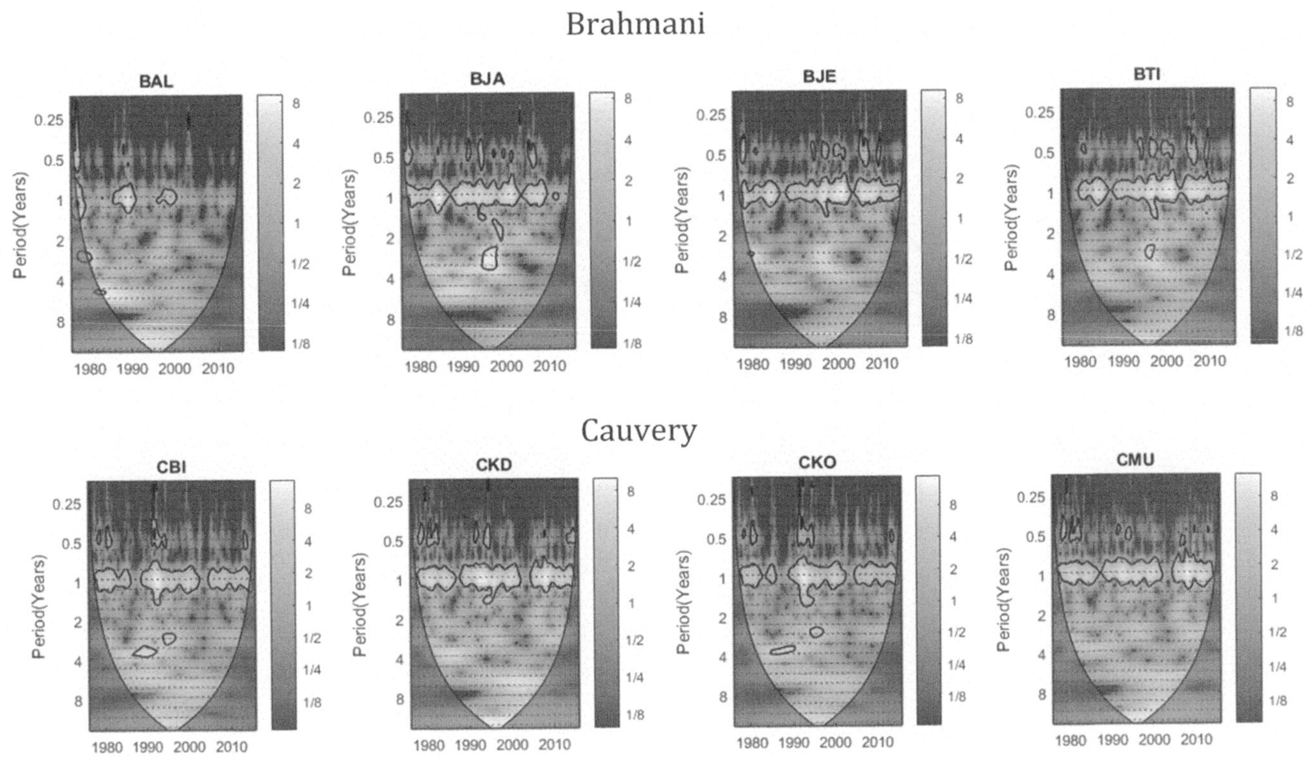

Figure 8 | Cross-wavelet analysis of Nino3.4 and 40 regulated streamflow stations of Brahmani, Cauvery, Godavari, Mahanadi, Narmada, and Subarnarekha basins in the same order listed in Table 1. The cone of influence indicates the area affected by the boundary assumption. (*continued.*).

5. DISCUSSION

Variation in streamflow can profoundly impact the national or possibly global spatial-temporal structure of water availability (Pahl-Wostl 2019). Frequent severe droughts or flood conditions directly impact water availability and significantly impact India's ecology and socio-economic growth (Rajamani 2009). This study identified streamflow variability of unregulated and regulated stations in terms of monotonic and abrupt change, followed by the time-frequency analysis of streamflow data of both types of stations. Finally, hydroclimatic teleconnections of streamflow with four climate indices are investigated.

5.1. Streamflow variability

Based on the Mann–Kendall trend test, at a 5% significance level, only two unregulated stations from the Cauvery and Narmada basins revealed a significant long-term negative trend. In contrast, most regulated stations in the Brahmani, Cauvery, and Godavari river basins showed a significant negative trend. In contrast, a significant positive trend was observed in Mahanadi, Narmada, and Subarnarekha river basins. The above results related to the Godavari and Subarnarekha basin are concurrent to the findings of various studies. Das (2019) and Kuriqi *et al.* (2020) studied the streamflow of various stations in the Godavari basin and concluded that upstream gauging stations show no significant trend. Despite this, the majority of the downstream stations recorded a significant downward trend in streamflow data. Similarly, upstream gauging sites in the Subarnarekha basin showed no trend, but downward gauging sites showed an increasingly significant trend. In the case of the Mahanadi basin, the results are in congruence with the findings of Murthy *et al.* (2017) and Panda *et al.* (2013), in which authors concluded a significant decreasing trend in the streamflow data.

Several major and minor storage projects have existed in all the six basins considered as characterized by Jain *et al.* (2017). For example, in the Brahmani basin, some major and medium projects were developed in the 1960s, 1980s, and 1990s. Several medium-sized storage projects were constructed in the Cauvery basin throughout the 1930s and 1950s and a few in the 1970s. The Godavari basin covers more than 0.3 million square kilometres and consists of several medium-sized storage projects built during the 1980s and 1990s. Still, no significant storage projects have been done since 2001. In 1958, the Hirakud

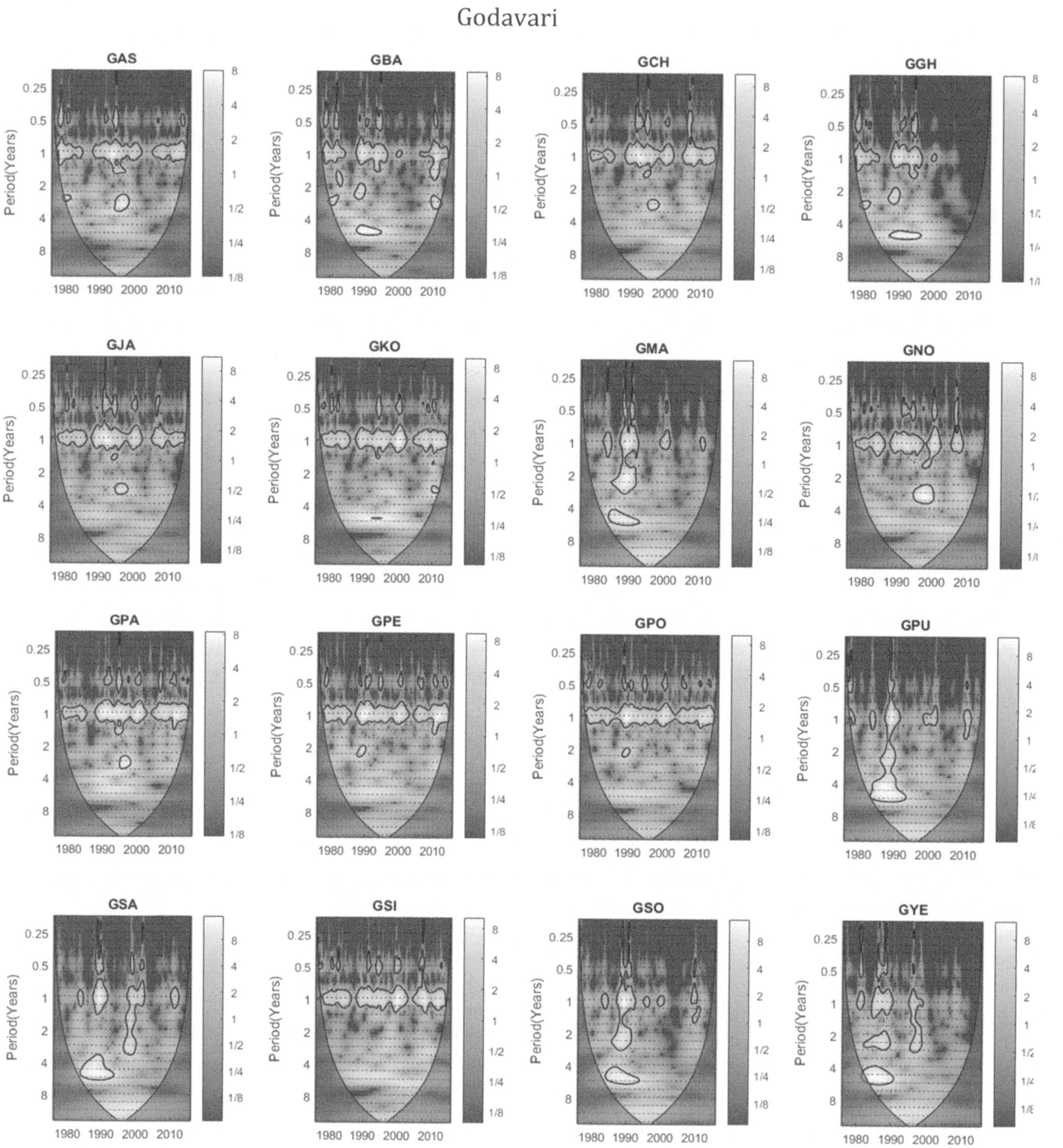

Figure 8 | Contined.

dam, a major project in the Mahanadi basin with a live storage capacity of 5.8 BCM, was completed. Following that, plenty of smaller projects proliferated. During the 1980s and 1990s, many medium-sized storage projects were established in the Narmada basin. Subarnarekha basin also contained several major and minor projects. Despite this, significant abrupt changes in regulated streamflow are much less common; on the other hand, all unregulated stations experienced a significant abrupt change in their streamflow data. Although unregulated catchments are not supposed to be influenced by hydropower plants, dams, or other reservoirs, such drastic changes in unregulated streamflow suggest that there may be other factors

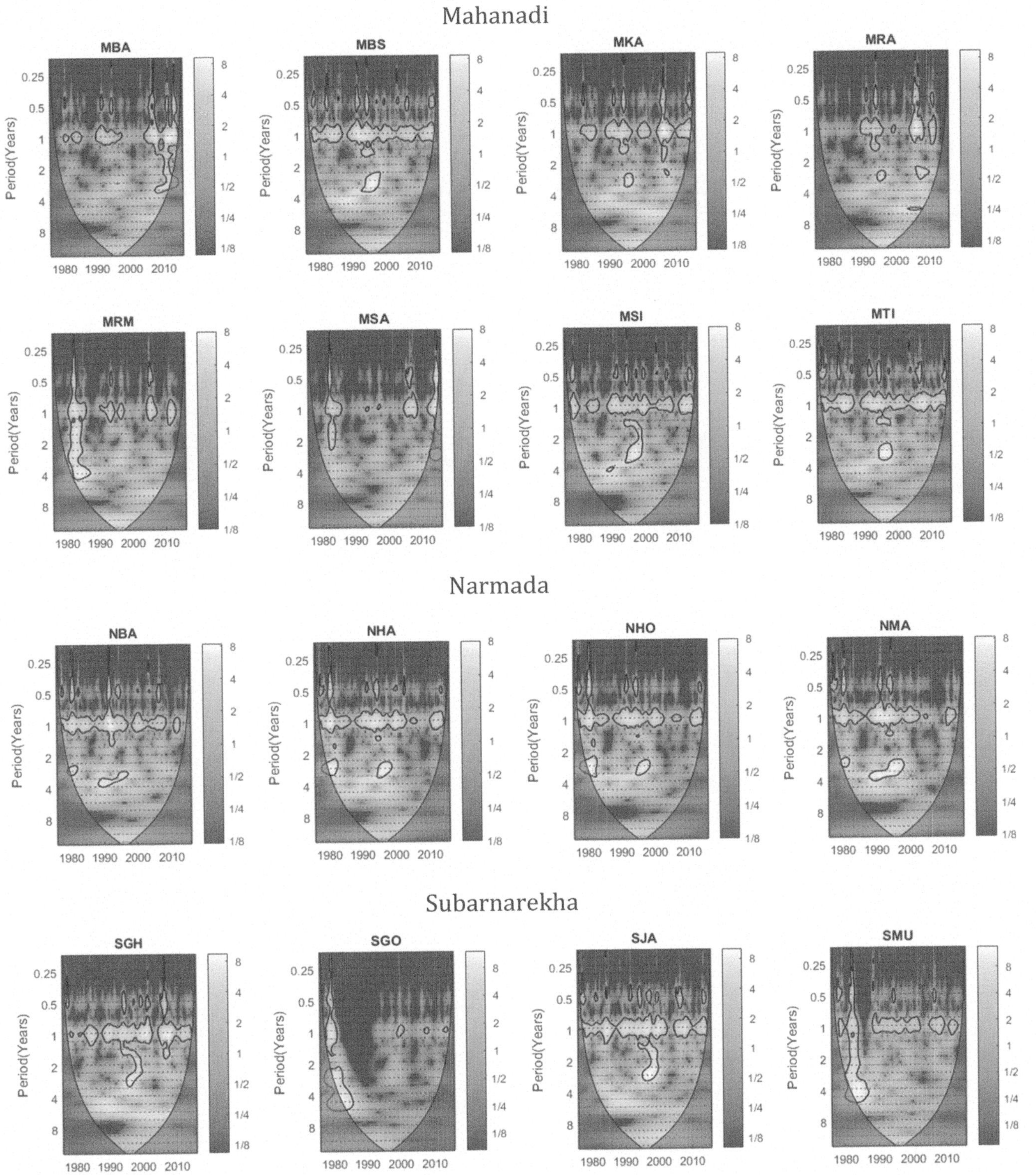

Figure 8 | Contined.

influencing variations in streamflow for these basins. According to Das *et al.* (2018), the main reason for the step-change in streamflow for the Subarnarekha River could be the strong influence of other nearby basins. Das *et al.* (2018) reported that urbanization and changes in land use land cover have a significant impact on streamflow alteration in addition to

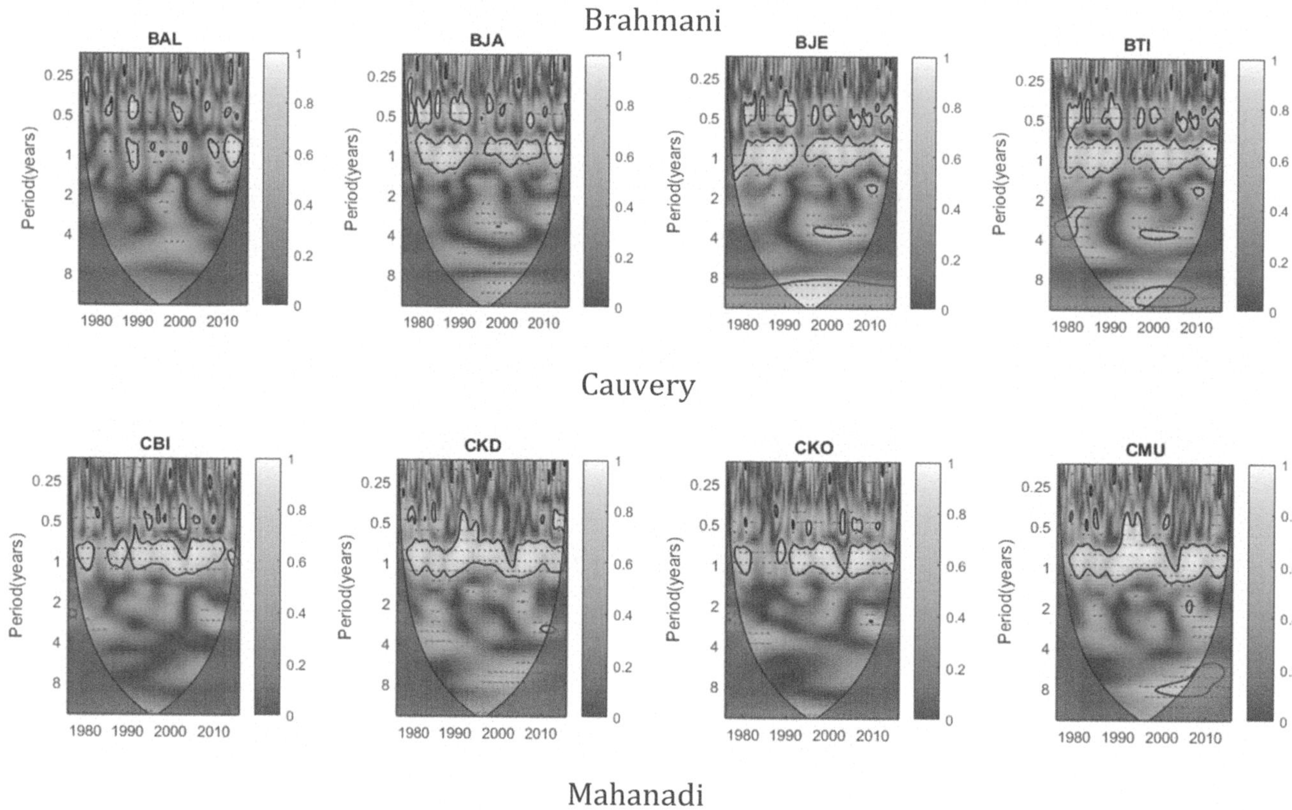

Figure 9 | Wavelet coherence analysis of Nino3.4 and 40 regulated streamflow stations of Brahmani, Cauvery, Godavari, Mahanadi, Narmada, and Subarnarekha basins in the same order listed in Table 1. The cone of influence indicates the area affected by the boundary assumption.

anthropogenic activity. Furthermore, they discussed increasing population density over these basins, causing cropland to be converted into built-up areas, causing surface water infiltration to be hampered, and increasing runoff.

CWT was used to understand better streamflow variability for the temporal analysis of streamflow data. All unregulated streamflow stations observed strong features at the intra-annual (0.5 years) and annual (1 year) time scales. Throughout the study period, the majority of the stations displayed a prominent feature at these scales. However, rather than intra-annual variability, oscillations from the annual to the intra-decadal scale dominate most regulated stations. Stations in regulated catchments observed a tiny region on a 0.5-year scale.

5.2. Hydroclimatic teleconnections

The IOD is essential for determining hydrologic factors such as streamflow and precipitation and acting as a modulator of the Indian monsoon (Ashok *et al.* 2001). This paper revealed that IOD had a minimal effect on streamflow in the basins studied, showing the spatial diversity of Indian streamflow teleconnections. This may be because IOD does not influence the hydrologic variables of the Indian basin compared with ENSO, which is also established by many studies (Ashok *et al.* 2003; Saji & Yamagata 2003; Marchant *et al.* 2007). In the absence of ENSO, Crétat *et al.* (2017) and Rathinasamy *et al.* (2019) investigated the impact of IOD on the Indian Summer Monsoon. They discovered that it did not push the monsoon circulation. Pokhrel *et al.* (2012) also found that the combined influence of ENSO and IOD improved forecasting of the Indian summer monsoon. It highlighted the necessity to investigate further these two indices' combined effect on streamflow variability in the Indian peninsular basin.

The essential areas of the vital feature in XWT and WTC are less with NAO than other indices, indicating less NAO dominance in these basins. A prior study by Yadav *et al.* (2009) supports this claim. The pressure in the polar region drops during the positive phase of NAO. In contrast, the pressure in the mid-latitudes (Indian subcontinent) rises. This positive phase of the NAO strengthens the western disturbance by intensifying the westerly jet stream over the Middle East (Nageswararao *et al.*

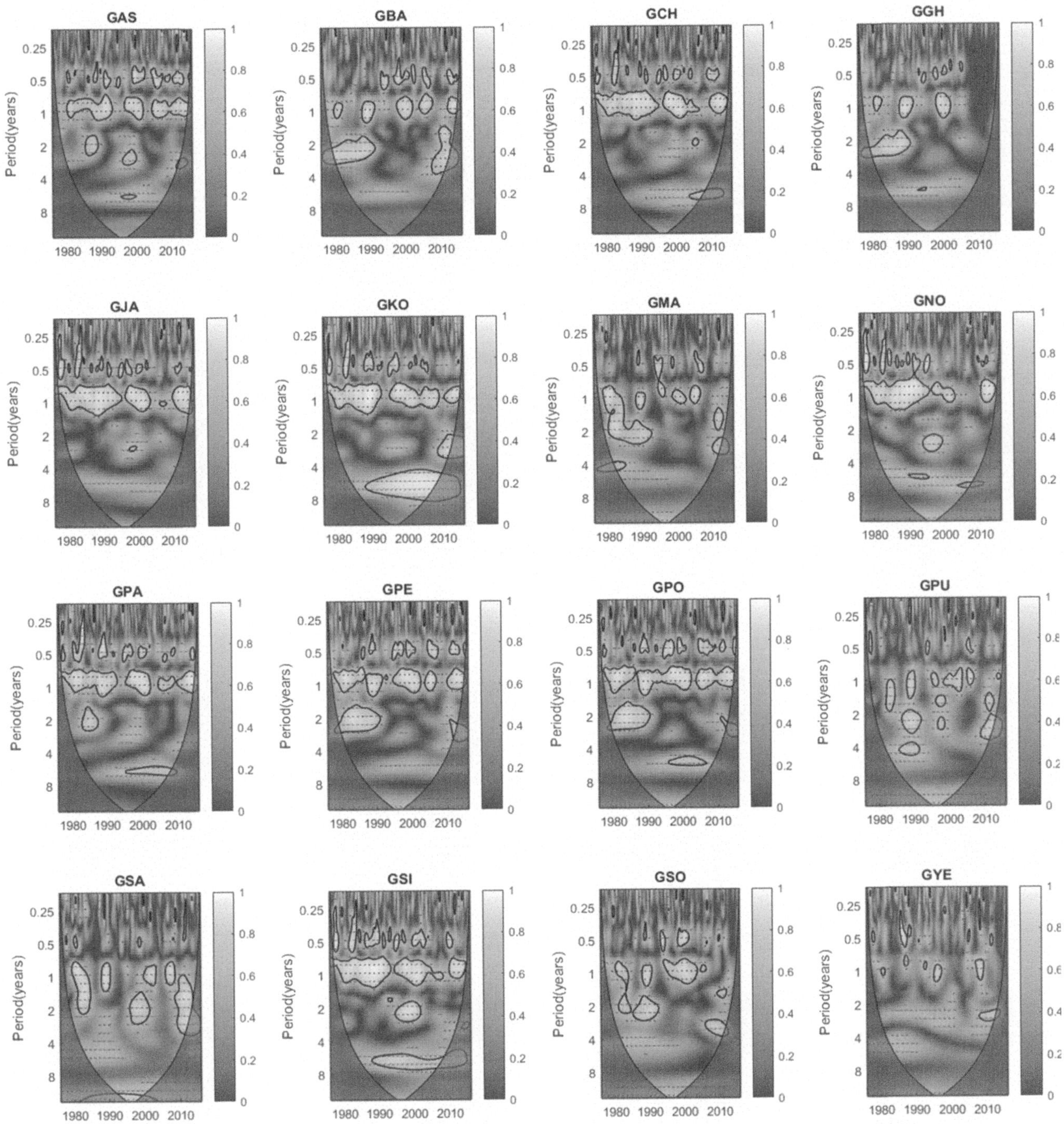

Figure 9 | Continued.

2016). Heavy rainfall in the winter season amplifies these disturbances, primarily impacting the northwest portion of the Indian subcontinent. Our research area is centred on the peninsular region, which has experienced less western disturbance and is unaffected by the NAO. Kurths *et al.* (2019) also found a weaker relationship between NAO and hydrologic factors in India's southern area.

Hydroclimatic teleconnections between streamflow and four climate indices unravel maximum significant correlation with Niño 3.4, followed by the PDO index. PDO influences on six basins were similar to Niño 3.4, highlighting the possibility of proxy connections (Rathinasamy *et al.* 2019). The relationship between PDO indices and Indian River basins is still fully

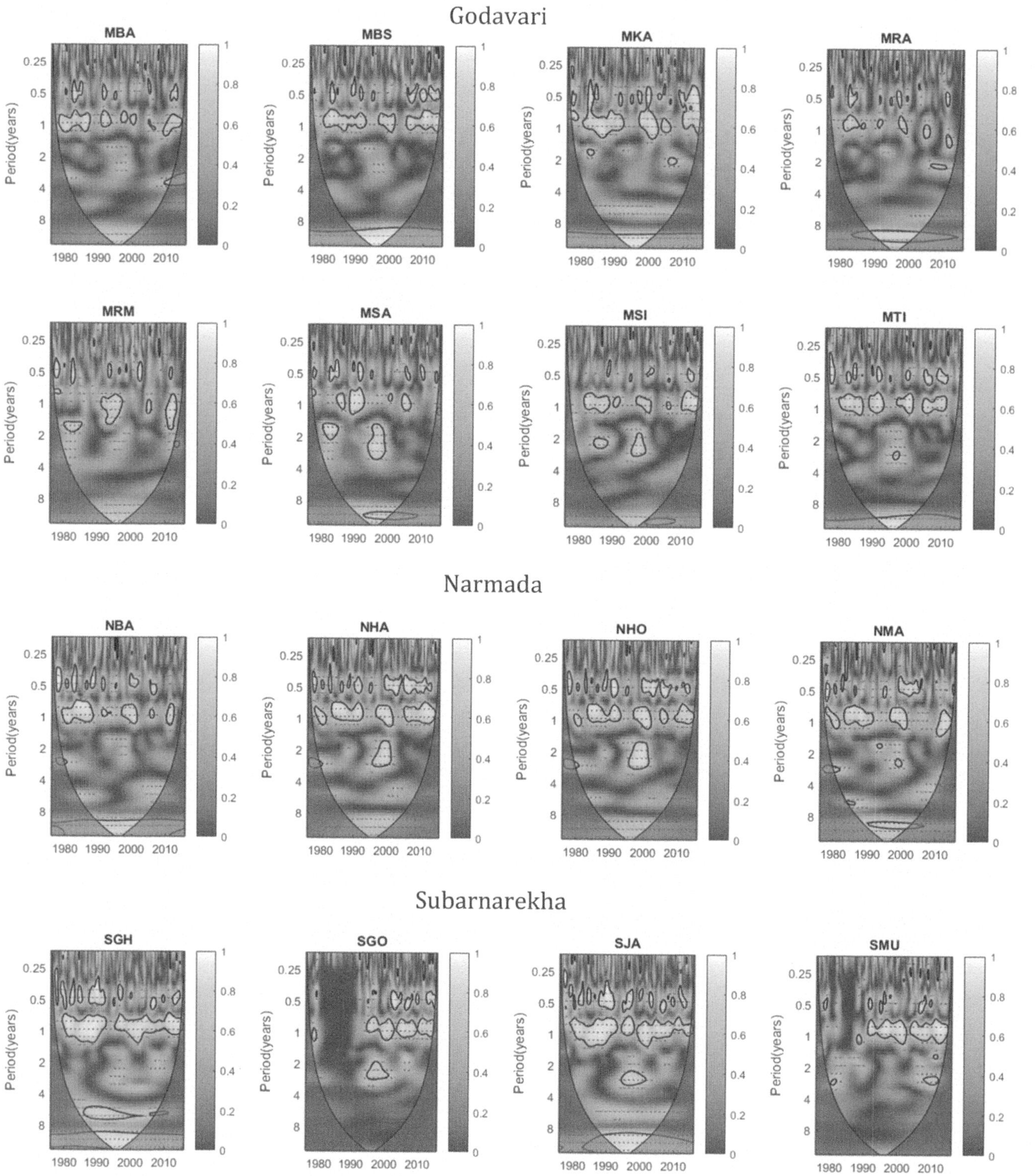

Figure 9 | Continued.

explored, except for a study by Krishnamurthy & Krishnamurthy (2014), who found that the warm and cold phases of PDO influence PDO ENSO conditions, affecting streamflow and precipitation in Indian basins. This could be related to the warm phase of the PDO, which raises the Pacific Ocean's sea surface temperature, altering trade winds and boosting the Walker Circulation, amplifying the Niño state (Agarwal 2019). Usually, moisture-driven currents move towards land due to pressure and temperature differences between the oceanic surface and the landmass of the Indian subcontinent. Still, during the Niño3.4 condition, the warm phase of ENSO, SSTs rise. As a result, the temperature difference between landmass and sea decreases, further hindering adequate moisture transport. A number of studies (Maity *et al.* 2007; Panda *et al.* 2013; Jena *et al.* 2014) established a positive relationship between streamflow and precipitation, implying that indices like PDO (positive phase) and Niño3.4 both have a negative impact on summer monsoon rainfall and thus reduce streamflow in the study area.

5.3. Comparison between regulated and unregulated streamflow variability and their hydroclimatic teleconnections

In unregulated catchments, all considered stations showed high variability at intra-annual, annual, and inter-annual scales. Oscillations at 0.5 and 1-year scales are more dominant and appeared throughout the study for almost all the stations. After the 1990s, the significant areas with oscillations at 2–4 years of scale can also be observed. But interestingly, for regulated stations rather than intra-annual variability, oscillations from annual to intra-decadal scale are dominant for most stations. Stations that belong to regulated catchments exhibit a less significant region at 0.5 years of scale. At annual and inter-annual scales, oscillations observed a similar pattern as unregulated, but regulated stations also observed some additional oscillation at the intra-decadal scale (8 years).

In hydroclimatic teleconnections, streamflow of unregulated stations is influenced mainly by climate indices at intra-annual, annual, and inter-annual scales. Mostly annual correlations are active between IOD and unregulated streamflow of Brahmani, Cauvery, Godavari, and Subarnarekha basin, but for Mahanadi and Narmada, inter-annual correlations can also be observed. With NAO, intra-annual and annual correlations are predominant, but some significant areas with high inter-annual correlations are also present. Similarly, both Niño 3.4 and PDO influence the streamflow from intra-annual to inter-annual scale. However, by looking at cross-wavelet and wavelet coherence plots between climatic indices and regulated streamflow, along with smaller scales, significant dominant connections at the intra-decadal scale (8 years) can also be observed. The oscillations at a larger scale can be observed for almost all basins except Mahanadi and Narmada, which do not observe a significant correlation with IOD and Niño 3.4 at the intra-decadal scale; however, the same station is showing correlation with NAO and PDO at this particular scale.

Previously, many basin-wise studies for the Indian peninsular regions were conducted (Panda *et al.* 2013; Jain *et al.* 2017; Murthy *et al.* 2017; Das 2019). However, the majority of the studies focus solely on regulated stations. Some works are available for unregulated stations as well, but none provide detailed information. In this study, streamflow variability and its response to climate indices for regulated and unregulated stations were identified separately.

Interestingly, in the context of streamflow variability and hydroclimatic teleconnections, this study observed the distinct behaviour of unregulated and regulated streamflow stations. Overall, streamflow variability and hydroclimatic teleconnections can be observed at a lower scale or a higher frequency for unregulated stations, indicating that variations in streamflow are more frequent and generally occur from intra-annual to inter-annual scale. The global climate indices also showed a correlation with unregulated streamflow at these scales, since climate drivers, such as IOD, NAO, and Niño 3.4, are periodic and usually appear every 2–3 years. This conveys the impression that climatic phenomena primarily influence streamflow at unregulated stations.

On the other hand, streamflow variability at the regulated station does not observe the significant oscillations at 0.5 years of scale, rather than the variation at a larger scale or lower frequencies (8–10 years). The hydroclimatic teleconnections also showed the correlation up to intra-decadal and decadal scale for most stations, which indicates that the variation in streamflow in regulated stations is not as frequent as unregulated ones. It also gives an idea that along with climatic patterns, some other factors, such as land use, land cover, population density, and other anthropogenic activity, are also predominately influenced by the streamflow at regulated stations, as suggested by Das *et al.* (2018).

By considering all these results obtained from both unregulated and regulated stations, the study concludes that both types of stations respond differently in any particular basin. So, the regulated station cannot be considered as a proxy of unregulated stations. Thus, for better planning and management of water resources, both regulated and unregulated streamflow should be examined. However, understanding the streamflow variability at any unregulated station provides a comparatively better understanding because, in the regulated station, regulated hydropower plants, dams, or any different structure alter the natural flow regime and affect the streamflow variability (Milner *et al.* 2019). The unregulated catchment, on the other hand, is unaffected by such changes. These water resource structures, such as dams or reservoirs, are typically built near the unregulated catchment's downstream end. Thus, unregulated stations, once again, provide a genuine image of any catchment and are useful for various practical applications like drainage structure design, runoff forecasts, and water distribution at the downstream end (Swain & Patra 2017).

6. CONCLUSION

This study attempted to investigate the nature of runoff variability among six different catchments of India and analyzed the influence of climate teleconnections on runoff variability. The study was carried out for six unregulated stations and 40 regulated stations from these six chosen basins. Mann–Kendall and step-change analysis was carried out to detect the gradual long-term and abrupt changes in the streamflow time series. Wavelet analysis was used to understand the multi-scale association of basin streamflow with the precipitation and climatic indices. The summary of significant findings from the study is as follows:

1. Based on the Mann–Kendall test and step-change analysis results, only two unregulated Narmada and Cauvery stations showed significant adverse changes in trend. In contrast, most regulated stations in the Brahmani, Cauvery, and Godavari river basins showed a significant negative trend. In contrast, a significant positive trend was observed in Mahanadi, Narmada, and Subarnarekha river basins. All unregulated stations experienced a significant abrupt change in their streamflow data. On the other hand, significant abrupt changes in regulated streamflow are much less common.
2. All CWT spectra of unregulated streamflow consistently show dominant periodicities at 0.5 and 1 year of band. In contrast, regulated stations are also observed streamflow variability at intra-decadal (8 years of scale), which suggests that higher (lower) frequency components might affect the streamflow variability of unregulated (regulated) streamflow stations.
3. The cross-wavelet and wavelet coherence plot between streamflow and different climate indices reveal that maximum strong feature at lower and higher frequencies is observed with NIÑO 3.4 index. These plots also show an in-phase relationship at the starting year and a lagged relationship at the end year with streamflow. Similarly, IOD and NAO show an in-phase relationship, and PDO indices showed a leading relationship with streamflow. All the global climate indices observed a relationship with unregulated streamflow up to inter-annual scale, whereas it can be observed up to intra-decadal scale for the regulated station.
4. Finally, the findings showed that both unregulated and regulated stations act differently in the context of streamflow variability and hydroclimatic teleconnections in any given basin. As a result, regulated stations cannot be used as a substitute for unregulated stations. Thus, both stations should be studied for improved water resource planning and management of any basin.

ACKNOWLEDGEMENTS

AA, AG, and TP acknowledge the funding support provided by the Indian Institute of Technology Roorkee through Faculty Initiation Grant number IITR/SRIC/1808/F.I.G and COPREPARE project funded by UGC and DAAD under the IGP 2020–2024.

COMPETING INTERESTS

The authors declare that they have no conflict of interest.

DATA AVAILABILITY STATEMENT

All relevant data are included in the paper or its Supplementary Information.

REFERENCES

Agarwal, A. 2019 Aufklärung raumzeitlicher Klimamuster über komplexe Netzwerke mit mehreren MaßstäbenUnraveling spatio-temporal climatic patterns via multi-scale complex networks. *Universität Potsdam*. https://doi.org/10.25932/PUBLISHUP-42395.

Ahmad, I., Tang, D., Wang, T., Wang, M. & Wagan, B. 2015 Precipitation trends over time using Mann-Kendall and Spearman's rho tests in Swat River Basin, Pakistan. *Adv. Meteorol.* **2015**, 1–15. https://doi.org/10.1155/2015/431860.

Alhaji, U. U., Yusuf, A. S., Edet, C. O., Oche, C. O. & Agbo, E. P. 2018 Trend analysis of temperature in Gombe State using Mann Kendall Trend Test. *J. Sci. Res. Rep.* **20**, 1–9. https://doi.org/10.9734/JSRR/2018/42029.

Arfan, M., Lund, J., Hassan, D., Saleem, M. & Ahmad, A. 2019 Assessment of spatial and temporal flow variability of the Indus River. *Resources* **8**, 103. https://doi.org/10.3390/resources8020103.

Ashok, K., Guan, Z. & Yamagata, T. 2001 Impact of the Indian Ocean dipole on the relationship between the Indian monsoon rainfall and ENSO. *Geophys. Res. Lett.* **28**, 4499–4502. https://doi.org/10.1029/2001GL013294.

Ashok, K., Guan, Z. & Yamagata, T. 2003 Influence of the Indian Ocean Dipole on the Australian winter rainfall. *Geophys. Res. Lett.* **30**. https://doi.org/10.1029/2003GL017926.

Beaupré, L., St-Hilaire, A., Daigle, A. & Bergeron, N. 2020 Comparison of a deterministic and statistical approach for the prediction of thermal indices in regulated and unregulated river reaches: case study of the Fourchue River (Québec, Canada). *Water Qual. Res. J.* **55**, 394–408. https://doi.org/10.2166/wqrj.2020.001.

Behera, S. K., Luo, J. J., Masson, S., Rao, S. A., Sakuma, H. & Yamagata, T. 2006 A CGCM study on the interaction between IOD and ENSO. *J. Clim.* **19**, 1688–1705. https://doi.org/10.1175/JCLI3797.1.

Birsan, M.-V., Zaharia, L., Chendes, V. & Branescu, E. 2014 Seasonal trends in Romanian streamflow. *Hydrol. Process.* **28**, 4496–4505. https://doi.org/10.1002/hyp.9961.

Briciu, A.-E. & Mihăilă, D. 2014 Wavelet analysis of some rivers in SE Europe and selected climate indices. *Environ. Monit. Assess.* **186**, 6263–6286. https://doi.org/10.1007/s10661-014-3853-z.

Burn, D. H. & Hag Elnur, M. A. 2002 Detection of hydrologic trends and variability. *J. Hydrol.* **255**, 107–122. https://doi.org/10.1016/S0022-1694(01)00514-5.

Chowdhury, M. R. & Ward, N. 2004 Hydro-meteorological variability in the greater Ganges–Brahmaputra–Meghna basins. *Int. J. Climatol.* **24**, 1495–1508. https://doi.org/10.1002/joc.1076.

Cohen, L. 1995 Uncertainty principles of the short-time Fourier transform. In: *Presented at the SPIE's 1995 International Symposium on Optical Science, Engineering, and Instrumentation* (Luk, F. T., ed.). San Diego, CA, pp. 80–90. https://doi.org/10.1117/12.211427

Coulibaly, P. & Burn, D. H. 2004 Wavelet analysis of variability in annual Canadian streamflows. *Water Resour. Res.* **40**. https://doi.org/10.1029/2003WR002667.

Crétat, J., Terray, P., Masson, S., Sooraj, K. P. & Roxy, M. K. 2017 Indian Ocean and Indian summer monsoon: relationships without ENSO in ocean–atmosphere coupled simulations. *Clim. Dyn.* **49**, 1429–1448. https://doi.org/10.1007/s00382-016-3387-x.

D'Arrigo, R. & Wilson, R. 2006 On the Asian expression of the PDO. *Int. J. Climatol.* **26**, 1607–1617. https://doi.org/10.1002/joc.1326.

Das, S. 2019 Four decades of water and sediment discharge records in Subarnarekha and Burhabalang basins: an approach towards trend analysis and abrupt change detection. *Sustainable Water Resour. Manage.* **5**, 1665–1676. https://doi.org/10.1007/s40899-019-00326-1.

Das, P., Behera, M. D., Patidar, N., Sahoo, B., Tripathi, P., Behera, P. R., Srivastava, S. K., Roy, P. S., Thakur, P., Agrawal, S. P. & Krishnamurthy, Y. V. N. 2018 Impact of LULC change on the runoff, base flow and evapotranspiration dynamics in eastern Indian river basins during 1985–2005 using variable infiltration capacity approach. *J. Earth Syst. Sci.* **127**, 19. https://doi.org/10.1007/s12040-018-0921-8.

Daubechies, I. 1990 The wavelet transform, time-frequency localization and signal analysis. *IEEE Trans. Inf. Theory* **36**, 961–1005. https://doi.org/10.1109/18.57199.

Drissia, T. K., Jothiprakash, V. & Anitha, A. B. 2019 Statistical classification of streamflow based on flow variability in west flowing rivers of Kerala, India. *Theor. Appl. Climatol.* **137**, 1643–1658. https://doi.org/10.1007/s00704-018-2677-0.

Ducré-Robitaille, J.-F., Vincent, L. A. & Boulet, G. 2003 Comparison of techniques for detection of discontinuities in temperature series. *Int. J. Climatol.* **23**, 1087–1101. https://doi.org/10.1002/joc.924.

Ehsanzadeh, E., Ouarda, T. B. M. J. & Saley, H. M. 2011 A simultaneous analysis of gradual and abrupt changes in Canadian low streamflows. *Hydrol. Process.* **25**, 727–739. https://doi.org/10.1002/hyp.7861.

Gadgil, S., Vinayachandran, P. N. & Francis, P. A. 2003 Droughts of the Indian summer monsoon: role of clouds over the Indian Ocean. *Curr. Sci.* **85**, 1713–1719.

Gocic, M. & Trajkovic, S. 2013 Analysis of changes in meteorological variables using Mann-Kendall and Sen's slope estimator statistical tests in Serbia. *Glob. Planet. Change* **100**, 172–182. https://doi.org/10.1016/j.gloplacha.2012.10.014.

Grinsted, A., Moore, J. C. & Jevrejeva, S. 2004 Application of the cross wavelet transform and wavelet coherence to geophysical time series. *Nonlinear Process. Geophys.* **11**, 561–566. https://doi.org/10.5194/npg-11-561-2004.

Growns, J. & Marsh, N. & Cooperative Research Centre for Freshwater Ecology (Australia) 2000 *Characterization of Flow in Regulated and Unregulated Streams in Eastern Australia*. CRC for Freshwater Ecology, Albury, NSW.

Guan, Z. & Yamagata, T. 2003 The unusual summer of 1994 in East Asia: IOD teleconnections. *Geophys. Res. Lett.* **30**, n/a–n/a. https://doi.org/10.1029/2002GL016831.

Gudmundsson, L., Leonard, M., Do, H. X., Westra, S. & Seneviratne, S. I. 2019 Observed trends in global indicators of mean and extreme streamflow. *Geophys. Res. Lett.* **46**, 756–766. https://doi.org/10.1029/2018GL079725.

Guntu, R. K., Rathinasamy, M., Agarwal, A. & Sivakumar, B. 2020 Spatiotemporal variability of Indian rainfall using multi-scale entropy. *J. Hydrol.* **587**, 124916. https://doi.org/10.1016/j.jhydrol.2020.124916.

Hamed, K. H. 2009 Enhancing the effectiveness of prewhitening in trend analysis of hydrologic data. *J. Hydrol.* **368**, 143–155. https://doi.org/10.1016/j.jhydrol.2009.01.040.

Hurrell, J. W. 2001 CLIMATE: the North Atlantic Oscillation. *Science* **291**, 603–605. https://doi.org/10.1126/science.1058761.

Immerzeel, W. W., van Beek, L. P. H. & Bierkens, M. F. P. 2010 Climate change will affect the Asian water towers. *Science* **328**, 1382–1385. https://doi.org/10.1126/science.1183188.

Itoh, N. & Kurths, J. 2010 *Change-Point Detection of Climate Time Series by Nonparametric Method 4*.

Jain, S. K., Nayak, P. C., Singh, Y. & Chandniha, S. K. 2017 Trends in rainfall and peak flows for some River Basins in India. *Curr. Sci.* **112**, 1712. https://doi.org/10.18520/cs/v112/i08/1712-1726.

Jena, P. P., Chatterjee, C., Pradhan, G. & Mishra, A. 2014 Are recent frequent high floods in Mahanadi basin in eastern India due to increase in extreme rainfalls? *J. Hydrol.* **517**, 847–862. https://doi.org/10.1016/j.jhydrol.2014.06.021.

Jiang, R., Gan, T. Y., Xie, J. & Wang, N. 2013 Spatiotemporal variability of Alberta's seasonal precipitation, their teleconnection with large-scale climate anomalies and sea surface temperature. *Int. J. Climatol.* n/a–n/a. https://doi.org/10.1002/joc.3883.

Joshi, N., Gupta, D., Suryavanshi, S., Adamowski, J. & Madramootoo, C. A. 2016 Analysis of trends and dominant periodicities in drought variables in India: a wavelet transform based approach. *Atmospheric Res.* **182**, 200–220. https://doi.org/10.1016/j.atmosres.2016.07.030.

Kawahara, Y. & Sugiyama, M. 2012 Sequential change-point detection based on direct density-ratio estimation. *Stat. Anal. Data Min.* **5**, 114–127. https://doi.org/10.1002/sam.10124.

Kisi, O. & Ay, M. 2014 Comparison of Mann–Kendall and innovative trend method for water quality parameters of the Kizilirmak River, Turkey. *J. Hydrol.* **513**, 362–375. https://doi.org/10.1016/j.jhydrol.2014.03.005.

Krishna Kumar, K. 1999 *On the Weakening Relationship Between the Indian Monsoon and ENSO | Science [WWW Document]*. Available from: https://science.sciencemag.org/content/284/5423/2156.abstract (accessed 7 February 2021).

Krishnamurthy, L. & Krishnamurthy, V. 2014 *Influence of PDO on South Asian Summer Monsoon and Monsoon–ENSO Relation 14.*

Kuriqi, A., Ali, R., Pham, Q. B., Montenegro Gambini, J., Gupta, V., Malik, A., Linh, N. T. T., Joshi, Y., Anh, D. T., Nam, V. T. & Dong, X. 2020 Seasonality shift and streamflow flow variability trends in central India. *Acta Geophys.* **68**, 1461–1475. https://doi.org/10.1007/s11600-020-00475-4.

Kurths, J., Agarwal, A., Shukla, R., Marwan, N., Rathinasamy, M., Caesar, L., Krishnan, R. & Merz, B. 2019 Unravelling the spatial diversity of Indian precipitation teleconnections via a non-linear multi-scale approach. *Nonlinear Process. Geophys.* **26**, 251–266. https://doi.org/10.5194/npg-26-251-2019.

Labat, D. 2010 Cross wavelet analyses of annual continental freshwater discharge and selected climate indices. *J. Hydrol.* **385**, 269–278. https://doi.org/10.1016/j.jhydrol.2010.02.029.

Lauro, C., Vich, A. I. J. & Moreiras, S. M. 2019 Streamflow variability and its relationship with climate indices in western rivers of Argentina. *Hydrol. Sci. J.* **64**, 607–619. https://doi.org/10.1080/02626667.2019.1594820.

Maheswaran, R. & Khosa, R. 2012 Comparative study of different wavelets for hydrologic forecasting. *Comput. Geosci.* **46**, 284–295. https://doi.org/10.1016/j.cageo.2011.12.015.

Maity, R. & Kashid, S. S. 2011 Importance analysis of local and global climate inputs for basin-scale streamflow prediction. *Water Resour. Res.* **47**. https://doi.org/10.1029/2010WR009742.

Maity, R. & Nagesh Kumar, D. 2008 Basin-scale streamflow forecasting using the information of large-scale atmospheric circulation phenomena. *Hydrol. Process.* **22**, 643–650. https://doi.org/10.1002/hyp.6630.

Maity, R., Kumar, D. N. & Nanjundiah, R. S. 2007 Review of hydroclimatic teleconnection between hydrologic variables and large-scale atmospheric circulation patterns with Indian perspective. *ISH J. Hydraul. Eng.* **13**, 77–92. https://doi.org/10.1080/09715010.2007.10514859.

Mantua, N. J. & Hare, S. R. 2002 The pacific decadal oscillation. *J. Oceanogr.* **58**, 35–44. https://doi.org/10.1023/A:1015820616384.

Marchant, R., Mumbi, C., Behera, S. & Yamagata, T. 2007 The Indian ocean dipole? The unsung driver of climatic variability in East Africa. *Afr. J. Ecol.* **45**, 4–16. https://doi.org/10.1111/j.1365-2028.2006.00707.x.

McLaughlin, F., Lapointe, M., Bourque, G. & Boisclair, D. 2014 Using regional flow classes as references to analyze flow regime anomalies across a set of regulated Canadian rivers. *J. Hydrol.* **519**, 307–328. https://doi.org/10.1016/j.jhydrol.2014.07.015.

Milner, V. S., Yarnell, S. M. & Peek, R. A. 2019 The ecological importance of unregulated tributaries to macroinvertebrate diversity and community composition in a regulated river. *Hydrobiologia* **829**, 291–305. https://doi.org/10.1007/s10750-018-3840-4.

Murthy, D. S., Jyothy, S. A. & Mallikarjuna, P. 2017 Trend analysis of annual maximum daily streamflows at the gauging sites of Godavari River Basin. *Hydrology Journal* **40 & 41** (1–4), 52–57.

Nageswararao, M. M., Mohanty, U. C., Ramakrishna, S. S. V. S., Nair, A. & Prasad, S. K. 2016 Characteristics of winter precipitation over Northwest India using high-resolution gridded dataset (1901–2013). *Glob. Planet. Change* **147**, 67–85. https://doi.org/10.1016/j.gloplacha.2016.10.017.

Neelin, J. D., Battisti, D. S., Hirst, A. C., Jin, F.-F., Wakata, Y., Yamagata, T. & Zebiak, S. E. 1998 ENSO theory. *J. Geophys. Res. Oceans* **103**, 14261–14290. https://doi.org/10.1029/97JC03424.

Pahl-Wostl, C. 2019 Governance of the water-energy-food security nexus: a multi-level coordination challenge. *Environ. Sci. Policy* **92**, 356–367. https://doi.org/10.1016/j.envsci.2017.07.017.

Panda, D. K., Kumar, A., Ghosh, S. & Mohanty, R. K. 2013 Streamflow trends in the Mahanadi River basin (India): linkages to tropical climate variability. *J. Hydrol.* **495**, 135–149. https://doi.org/10.1016/j.jhydrol.2013.04.054.

Petpongpan, C., Ekkawatpanit, C. & Kositgittiwong, D. 2020 Climate change impact on surface water and groundwater recharge in Northern Thailand. *Water* **12**, 1029. https://doi.org/10.3390/w12041029.

Pokhrel, S., Chaudhari, H. S., Saha, S. K., Dhakate, A., Yadav, R. K., Salunke, K., Mahapatra, S. & Rao, S. A. 2012 ENSO, IOD and Indian Summer Monsoon in NCEP climate forecast system. *Clim. Dyn.* **39**, 2143–2165. https://doi.org/10.1007/s00382-012-1349-5.

Radke, R. J., Andra, S., Al-Kofahi, O. & Roysam, B. 2005 Image change detection algorithms: a systematic survey. *IEEE Trans. Image Process. Publ. IEEE Signal Process. Soc.* **14**, 294–307. https://doi.org/10.1109/tip.2004.838698.

Rajamani, L. 2009 India and climate change: what India wants, needs, and needs to do. *India Rev.* **8**, 340–374. https://doi.org/10.1080/14736480903116842.

Rathinasamy, M., Agarwal, A., Sivakumar, B., Marwan, N. & Kurths, J. 2019 Wavelet analysis of precipitation extremes over India and teleconnections to climate indices. *Stoch. Environ. Res. Risk Assess.* **33**, 2053–2069. https://doi.org/10.1007/s00477-019-01738-3.

Rathnayake, U. 2019 Comparison of statistical methods to graphical methods in rainfall trend analysis: case studies from tropical catchments. *Adv. Meteorol.* **2019**, 1–10. https://doi.org/10.1155/2019/8603586.

Reeves, J., Chen, J., Wang, X. L., Lund, R. & Lu, Q. Q. 2007 A review and comparison of changepoint detection techniques for climate data. *J. Appl. Meteorol. Climatol.* **46**, 900–915. https://doi.org/10.1175/JAM2493.1.

Rochdane, S., Reichert, B., Messouli, M., Babqiqi, A. & Khebiza, M. Y. 2012 Climate change impacts on water supply and demand in Rheraya watershed (Morocco), with potential adaptation strategies. *Water* **4**, 28–44. https://doi.org/10.3390/w4010028.

Rybach, D., Gollan, C., Schluter, R. & Ney, H. 2009 Audio segmentation for speech recognition using segment features. In: *2009 IEEE International Conference on Acoustics, Speech and Signal Processing*. IEEE, Taipei, Taiwan, pp. 4197–4200. https://doi.org/10.1109/ICASSP.2009.4960554

Sadowsky, J. 1996 Investigation of signal characteristics using the continuous wavelet transform. *JOHNS HOPKINS APL Tech. Dig.* **17**, 12.

Sahoo, B. B. & Jha, R. 2020 Assessment of low flow trends and change point detection in Mahanadi River basin. *India. Sustain. Water Resour. Manage.* **6**, 81. https://doi.org/10.1007/s40899-020-00441-4.

Sahu, N., Behera, S. K., Yamashiki, Y., Takara, K. & Yamagata, T. 2012 IOD and ENSO impacts on the extreme streamflows of Citarum river in Indonesia. *Clim. Dyn.* **39**, 1673–1680. https://doi.org/10.1007/s00382-011-1158-2.

Saji, N. & Yamagata, T. 2003 Possible impacts of Indian Ocean Dipole mode events on global climate. *Clim. Res.* **25**, 151–169. https://doi.org/10.3354/cr025151.

Saji, N. H., Goswami, B. N., Vinayachandran, P. N. & Yamagata, T. 1999 A dipole mode in the tropical Indian Ocean. *Nature* **401**, 360–363. https://doi.org/10.1038/43854.

Schulte, J. A., Najjar, R. G. & Li, M. 2016 The influence of climate modes on streamflow in the Mid-Atlantic region of the United States. *J. Hydrol. Reg. Stud.* **5**, 80–99. https://doi.org/10.1016/j.ejrh.2015.11.003.

Shadmani, M., Marofi, S. & Roknian, M. 2012 Trend analysis in reference evapotranspiration using Mann-Kendall and Spearman's Rho Tests in Arid Regions of Iran. *Water Resour. Manag.* **26**, 211–224. https://doi.org/10.1007/s11269-011-9913-z.

Swain, J. B. & Patra, K. C. 2017 Streamflow estimation in ungauged catchments using regional flow duration curve: comparative study. *Journal of Hydrologic Engineering* **22**, 04017010.

Taylor, W. 2000 *Change-Point Analysis: A Powerful New Tool For Detecting Changes 19.*

Tran, D.-H. 2019 Automated change detection and reactive clustering in multivariate streaming data. In: *2019 IEEE-RIVF International Conference on Computing and Communication Technologies (RIVF)*. IEEE, Danang, Vietnam, pp. 1–6. https://doi.org/10.1109/RIVF.2019.8713738.

Visbeck, M. H., Hurrell, J. W., Polvani, L. & Cullen, H. M. 2001 The North Atlantic Oscillation: past, present, and future. *Proc. Natl. Acad. Sci.* **98**, 12876–12877. https://doi.org/10.1073/pnas.231391598.

Yadav, R. K., Rupa Kumar, K. & Rajeevan, M. 2009 Increasing influence of ENSO and decreasing influence of AO/NAO in the recent decades over northwest India winter precipitation. *J. Geophys. Res.* **114**, D12112. https://doi.org/10.1029/2008JD011318.

Yang, P., Dumont, G. & Ansermino, J. M. 2006 Adaptive change detection in heart rate trend monitoring in anesthetized children. *IEEE Trans. Biomed. Eng.* **53**, 2211–2219. https://doi.org/10.1109/TBME.2006.877107.

Yeditha, P. K., Pant, T., Rathinasamy, M. & Agarwal, A. 2021 Multi-scale investigation on streamflow temporal variability and its connection to global climate indices for unregulated rivers in India. *Journal of Water and Climate Change*, jwc2021189. doi: https://doi.org/10.2166/wcc.2021.189.

First received 27 September 2021; accepted in revised form 26 December 2021. Available online 24 January 2022

IWA Publishing's authorised EU representative for General Product Safety Regulations is Diane D'Arras, 15 rue Duret, 75116 Paris, France, e-mail: safety@iwap.co.uk.

Printed and bound by CPI Group (UK) Ltd, Croydon, CR0 4YY
16/06/2026
02140402-0001